D0630846

CALGARY PUBLIC LIBRARY

OCT   2013

Stalin's Legacy

Stalin's Legacy

The Soviet War on Nature

Struan Stevenson

BIRLINN

First published in 2012 by
Birlinn Limited
West Newington House
10 Newington Road
Edinburgh
EH9 1QS

www.birlinn.co.uk

Copyright © Struan Stevenson 2012
Foreword copyright © George Robertson 2012

All rights reserved.
No part of this publication may be reproduced,
stored or transmitted in any form without the express
written permission of the publisher.

ISBN: 978 1 78027 090 6

British Library Cataloguing-in-Publication Data
A catalogue record for this book is available from the British Library

Designed and typeset by Iolaire Typesetting, Newtonmore
Printed and bound by MPG Books Ltd, Bodmin

'Once the class struggle has been won, Soviet humankind will be free to engage its final enemy: nature.'

Maxim Gorky (1868–1936)

Contents

List of Illustrations

The first Soviet nuclear explosion, nicknamed 'Joe-1' by the Americans.

One of the many radioactive craters that have ruined farmland and water bodies in the Polygon.

A boy born to a Soviet pilot and his wife who were working on the nuclear tests in the Polygon, with a single eye in the centre of his forehead – a perfect Cyclops.

Some of the citizens of the Polygon have suffered terrible deformities due to the legacy of nuclear testing.

Angry Kazakh villagers from Sarzhal in the Polygon have called for more help for victims of the Soviet nuclear tests.

The Tsar Bomba – the largest, most powerful nuclear weapon ever detonated.

Struan Stevenson MEP, and UN Secretary General Ban Ki-Moon during a visit to Kurchatov.

Struan Stevenson Street in Znamenka.

Struan Stevenson with parents and children at the opening of the new Urdzhar School for Handicapped Children.

The rapidly receding shoreline of the Aral Sea has left many fishing boats strewn across the former seabed.

Struan Stevenson on one of the rotting hulks at Muynak harbour, which is now over 100km from the sea.

Changes in the water level of the Aral Sea.

With the wife of the *akim* of Muynak after being welcomed into the furnace-like interior of their home and naming their grandson 'William Wallace'.

The Nurek dam in Tajikistan, currently the tallest in the world.

Struan Stevenson at the Nurek dam in Tajikistan, with the slogan 'Water is Life' painted over a tunnel entrance.

Struan Stevenson with Tajik president Emomali Rahmon at the Rogun dam.

Struan Stevenson with one of the less aggressive fighting dogs near Ashkhabad in Turkmenistan.

Maps

Acknowledgements

My sincere thanks to Ben Acheson, who devoted much of his holidays and free time over a period of 18 months to researching, checking and correcting facts and suggesting improvements to this book; his invaluable insights and meticulous research have been an inspiration.

My great thanks also to the friends, advisors, collaborators, interpreters and co-conspirators who have accompanied me on many exciting, exhausting and sometimes dangerous expeditions to Central Asia and Russia: Kamila and Sulushash Magzieva, Elena Kachkova, Anna Dmitrijewa and Kimberley Joseph. *Stalin's Legacy* would not have happened without you.

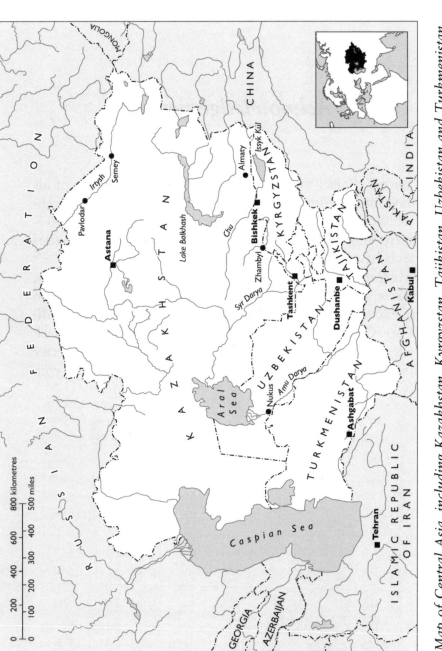

Map of Central Asia, including Kazakhstan, Kyrgyzstan, Tajikistan, Uzbekistan and Turkmenistan.
(Source: adapted from UN Map 37363 Rev. 7)

Foreword

This is a modern horror story about a ruthless regime using its own people like rats in an experiment. Told in very personal style by a campaigning Member of the European Parliament, the appalling legacy of what was Stalin's Soviet Union is spelt out in graphic terms. It makes for uncomfortable but compelling reading.

At one of my early meetings as NATO Secretary General with the then new President Putin, he expressed his concern with the proliferation of nuclear and other lethal material. He was candid about his country's record. 'Many things happened as the Soviet Union broke up. We are still not able to say how much technology and material escaped or was sold.' He saw proliferation as a key subject for NATO/Russia cooperation, and so did we.

The fact that the Soviet Union had huge capabilities in the nuclear, chemical and biological warfare field was never a secret. Indeed it was one of their boasts. What was unappreciated by everyone except a small Soviet elite was the brutal, merciless way in which Stalin had developed these capabilities using his own people and the land they lived on as a test-bed. Today the people of these lands still suffer the agonies left by a man completely careless of the humanity he abused.

Anyone who flies over modern Kazakhstan, as I have done, is struck by the mysterious straight lines criss-crossing thousands of miles on the ground. The observer will also see the Aral Sea, once a mighty internal ocean, now reduced to a miserable puddle surrounded by the concentric circles measuring the retreat of the water. In this book the mystery is dispelled. The perpetrator was not nature, it was an evil tyrant who used the periphery of his empire as a gigantic cruel laboratory.

Struan Stevenson not only takes us on a personal journey to explore and expose the scandal of Stalin's war on the environment, but he usefully, and importantly tells us what needs to be done to avoid such outrages happening again.

Rt Hon. Lord Robertson of Port Ellen
KT GCMG HonFRSE PC
Secretary General,
North Atlantic Treaty Organisation 1999–2003

Introduction

An atomic lake, an imploded mountain, a disappearing sea, a top-secret biological weapons-testing site and hundreds of millions of tonnes of radioactive waste; contaminated food, deformed babies and widespread illness. Welcome to Central Asia, and some of the world's greatest environmental disasters. As undisputed leader of the Soviet Union, Joseph Vissariono-vich Stalin (18 December 1878 – 5 March 1953) introduced a policy of rapid industrialisation and the brutal collectivisation of agriculture that led to widespread famine and a catastrophic death toll. During the late 1930s, he launched the 'Great Terror', a campaign to purge the Communist Party of people accused of sabotage, terrorism or treachery. He extended it to the military and other sectors of Soviet society. In practice, the purges were indiscriminate; tens of thousands of innocent victims were executed, imprisoned in Gulag labour camps in Siberia and Central Asia or exiled. In the years which followed, millions of members of ethnic minorities were also deported. It is estimated that up to 60 million Soviet citizens lost their lives as a direct result of Stalin's repressive reign, making him one of the greatest butchers in history.

But Stalin not only waged war on his own people. He and some of his immediate successors regarded nature as an enemy that could be overcome by the might of Soviet technology and the brute force of slave labour. Stalin ordered vast networks of canals and irrigation channels to be dug by hand in an attempt to transform deserts into lush pastures. He built gigantic dams and reservoirs, and diverted the course of major rivers. He used his own citizens as human guinea pigs for nuclear tests, and he conducted top-secret biological weapons experiments on islands that had been cleared of all animal and insect life.

The legacy of Stalin's ill-considered environmental adventures has been devastating. In Central Asia, the Aral Sea has been virtually drained; toxic dust storms rage across the landscape; humans have been exposed to anthrax, typhus and other deadly bio-weapons; generations face illness and disease because of exposure to radiation; acute water shortages threaten regional conflict and the mass migration of environmental refugees. These are global issues with global consequences that may yet impact on all of us.

Since being elected to the European Parliament representing Scotland back in 1999, I have travelled extensively in Central Asia. Having discovered the horrific legacy of Soviet nuclear tests in the Polygon of East Kazakhstan, I have returned there many times. I wrote a book, *Crying Forever*, about my experiences, the sales of which have raised money which Mercy Corps has distributed to children's hospitals, cancer hospitals, village clinics and other important projects in the region.

To show their appreciation, the Kazakhs have given me the Freedom of the City of Semipalatinsk, the main city of East Kazakhstan and the administrative centre of the nuclear-testing zone during Soviet times. I was the first and only foreigner ever to receive this honour; they also decorated me with various medals and honorary professorships and doctorates. In 2010, when Kazakhstan took over the rotating presidency of the Organisation for Security and Cooperation in Europe (OSCE), they invited me to become their 'roving ambassador' or personal representative of the chairman in office, with the special remit to draw up a report on the environment of the five Central Asian republics. My travels around Central Asia and the almost unimaginable environmental horror stories I uncovered, all arising from Stalin's determination to conquer nature, are the subject of this book.

CHAPTER ONE

Central Asia

Kazakhstan is home to more than 100 different ethnic groups and 45 religions, all of which live at peace with one another. Indeed the country's leader, President Nursultan Nazarbayev, now promotes racial and religious harmony as one of Kazakhstan's greatest achievements. Nazarbayev himself was born on a collective farm, to scarcely literate parents descended from nomads, in the foothills of the Tien Shan Mountains. He joined the Communist Party in 1962, and as a steelworker earned a reputation for energy and leadership. He soon became first secretary of the Young Communists in his steel plant and steadily rose through the ranks of the Party until he was appointed secretary of the Central Committee of the Kazakh Communist Party in 1980. He became president of the newly independent republic after the collapse of the USSR.

Over half the population is of Kazakh origin, and Russians comprise just over a quarter of the population, with smaller minorities of Uzbeks, Koreans and Chechens accounting for the rest. Since gaining independence following the collapse of the Soviet Union in 1991, Kazakhstan has become a model of stability and prosperity in the region.

Kazakhstan is the ninth largest country in the world – bigger than Western Europe – but with a population of only 15.5 million people. It is rich in mineral resources and its oil reserves are said to be as big as those of Saudi Arabia, with exploitable deposits of coal, iron, lead, aluminium, zinc, uranium, silver and gold. Its landmass ranges from majestic mountains of Himalayan stature on the borders with Mongolia and China, to endless expanses of arid 'steppe', sparsely dotted by remote Kazakh villages. The country has more than 40,000 lakes, which are teeming with fish and home to immense flocks of flamingos. Its great rivers, like the Irtysh and the Ili, were once the watering places for the conquering hordes commanded by Genghis Khan.

Since independence, Kazakhstan has propelled itself into the premier league of economic tigers in Central Asia with year-on-year growth in excess of 10 per cent, despite a momentary slowing of the economy during the world recession. The EU is Kazakhstan's biggest trading partner, and the state is pursuing a strategy of advanced social, economic and political modernisation which is creating a positive environment for investors. Massive investment is going into new, secure oil and gas transit routes to the West, to ensure a steady supply of hydrocarbons to European consumers.

Despite its rising prosperity, millions of people worldwide only know Kazakhstan from the antics of Borat, the hapless TV reporter played by Sacha Baron-Cohen, who claims to be from Kazakhstan. Many people in the West encountered Kazakhstan for the first time in this comedy blockbuster movie, although none of the film was actually shot in Central Asia. In fact, most of the movie was shot in Romania, where the abject poverty of Kazakh villagers is portrayed as playing a central role in Borat's life story.

Borat may have been filmed in Romania, but despite its burgeoning economic growth, poverty is still a real problem in parts of Kazakhstan, particularly the remote villages of the steppe, where many villagers complain that they have been forgotten. Not only do these formerly nomadic people have to live with the aftermath of Soviet-era nuclear testing and toxic waste dumping, but they also have to contend with common social problems like increasing drug addiction and a growing incidence of HIV/Aids.

Elections in April 2011 handed a landslide victory to Nursultan Nazarbayev for a further seven-year term as president with more than 95 per cent of the vote. His grip on power was strengthened even further when parliament voted in 2007 to allow him to stay in office for an unlimited number of terms. Although he says he advocates democracy as a long-term goal, he warns that stability could be at risk if change is too swift. In 2010 the Majilis (the Kazakh parliament) agreed that there was no one capable of replacing Nazarbayev as president in the foreseeable future and passed a law requiring a referendum to be held that will enable him to remain in office until at least 2020. Nazarbayev wisely quashed this law, recognising that the 'Arab Spring' uprisings raging across the Middle East in 2011 often arose because rulers had clung to power for too many decades. He wishes to go down in history as the founding father of modern Kazakhstan and not as a ruler forced from power by a popular uprising. Nevertheless, his overwhelming victory in the 2011 presidential election will keep the 71-year-old in office until at least 2018.

The president merged his Otan party with his daughter Dariga's party, Asar, in July 2006. The move created a vast ruling coalition and was seen as consolidating the president's power. Otan was subsequently renamed Nur-Otan in honour of President Nazarbayev, and Dariga, who had been widely tipped as a possible successor, was quietly sidelined. There are currently no opposition MPs in the country's Majilis.

I have met President Nazarbayev several times. In 2005, I led a team of election observers from the European Parliament to cover the presidential elections. I was summoned for a personal audience to the grand new marble edifice of the presidential palace in the Kazahk capital, Astana, designed by British architect Lord Foster of Thamesbank (Norman Foster). I told Nazarbayev that he was a popular leader of his country and that independent opinion polls conducted by the Americans showed that he would win the election with a handsome majority. But, I warned, if the majority was around 60 or 65 per cent, no one in the West would raise an eyebrow. If it was 90 or 95 per cent, people would think that there had been a swindle.

The president roared with laughter. 'I can't help being popular,' he said. He then launched into a lecture on the emergence of democracy in Kazakhstan, explaining that following the collapse of the USSR he was determined to bring a new parliamentary system to his country, but that in the entire history of Central Asia there had never been such a thing as democracy, only strong leaders that the people looked up to, such as Genghis Khan or Timur (Tamerlane). He said he had literally gathered his closest political friends and allies together and told them: 'You, you and you will form the opposition and you, you and you will join me in the government. There were howls of protest from those I told to make up the opposition parties, but I had to explain to them that democracy cannot work without at least a two-party system and democratic elections.'

On that occasion I took my team of around ten Euro MPs and officials to Astana, and we visited countless polling stations and spoke to opposition politicians and the media. The Kazakhs were very proud of the fact that they had bought and installed a new electronic voting system, which we saw in many polling stations. The big joke amongst Kazakhs was that these complex machines had been built in Belarus, and they predicted that the authoritarian Belarusian President Alexander Lukashenko would be the ultimate winner of the Kazakh presidential elections!

Bruce George was the leader of the huge team of 465 election observers from OSCE. A large, bluff Welshman, Bruce was at the time the Labour Party member of parliament for Walsall in the House of Commons. He suggested that we should gather together all of our team leaders for a strictly confidential meeting following the closing of the polling stations, so that we could work out the main thrust of our communiqué for the press conference the next morning. He asked me where we could meet in Astana that was totally secure and well away from prying eyes and ears. We decided to approach the Americans to see if they would allow us to use a room in the US Embassy.

At around midnight, Bruce and his senior advisers from the OSCE, together with me and my top officials, arrived outside the US Embassy gates in a suburb of Astana. Extensive security checks were conducted on our car before the enormous tank barriers and iron gates were swung open. Once in the compound we were escorted to a lift and taken up to the seventh floor of the building, where we were shown into a large meeting room which, we were assured, had been carefully scanned for bugging devices. We set about exchanging views on the conduct of the elections.

Around two in the morning, someone said that none of the recommendations we had made following the Majilis elections in 1999 had been implemented. Bruce George said that he was reluctant to put this into our final communiqué unless we were absolutely certain of our facts. He instructed our officials to check carefully 'even if it takes all night' to see if any of the recommendations had actually made it onto the Kazakh statute books.

After that we all wearily made our way back to our hotels scattered across the city in temperatures that had dipped to around −20°C. I got to my hotel at about 3 a.m., and as I turned the key in my bedroom door, was astonished to find a large, unmarked, brown envelope lying on the floor. It had been slid under my door. I tore it open to find a comprehensive list of all the recommendations made by the OSCE following the 1999 Majilis elections and how each of them had been properly implemented into Kazakh law! Next morning, each of the team leaders and officials who had attended the high-security meeting in the US Embassy the previous evening said they had returned to their hotels to find similar envelopes!

It wasn't so much that every word we said was being listened to that concerned us, as the blatant way the Kazakh authorities were happy to let us know. President Nazarbayev was re-elected with 91 per cent of the popular vote. We reported that although there had been some improvements in the democratic process, there was still a long way to go. But perhaps we shouldn't be surprised at the Kazakhs' rather distorted version of democracy. After all, throughout their entire history, democracy has been an alien concept across the whole of Central Asia.

A Reflection on the History of Central Asia

In the many years I have visited Central Asia since the fall of the Soviet empire, I have witnessed dramatic changes and huge improvements to the standard of living of the population,

particularly in Kazakhstan. The first time I visited Semipalatinsk there were no street lights and few of the roads were
paved. After dark, you could see small fires flickering along the
roadsides, as people came out onto the streets to cook their
evening meals on charcoal burners. Few of the houses had
electricity, running water or proper sanitation.

Within ten years, that has changed. The streets are paved
and lit, the houses have been fitted with electricity, water
and sanitary facilities, and shops, hotels and businesses are
springing up in increasing abundance. In fact, traffic conges-
tion and air pollution are now visible signs of the growing
prosperity of these nations, although there is evidence of a
widening gap between the richest and poorest members of
society. The city streets of Astana and Almaty are clogged
with black four-wheel-drive Porsche Cayennes with darkened
windows, which conceal the growing number of wealthy,
middle-class entrepreneurs that this capitalist society has
spawned.

But scratch the surface and you can find a bit of Central Asia
in most of us. It is reckoned that Homo sapiens first settled in
the region 40,000 to 50,000 years ago, making it one of the
earliest sites of human habitation. Some other studies have
shown that Central Asia was the probable source for the
nomadic peoples who later spread to Europe, Siberia and
North America. This means that many of us are directly
descended from Central Asians. Experts also believe that
our common Indo-European languages had their roots in
Central Asia.

Six and a half thousand years ago, small family groups of
these nomadic peoples began to form tiny communities, which
in turn began to build permanent settlements. With the need to
increase their supply of food beyond that which could be
obtained simply by hunting, they began to practise farming
and the herding of livestock. Animals that were the forerunners
of modern camels, cows, pigs, sheep and goats were domes-
ticated, including, around the same time, horses, which had
previously been bred only for meat. In due course, wheeled
carts and chariots appeared, and gradually horses were bred to

be bigger and stronger, until enormous warhorses, capable of carrying heavily armed warriors, evolved.

The nomadic tribes who now tended herds of camels and cows, and flocks of sheep and goats, constantly roamed across the Central Asian steppe in search of fresh pastures for grazing and water for their animals. These nomadic tribes developed a transportable shelter system in the form of yurts – tent-like structures made of wool – which could be erected quickly around wooden staves and could accommodate entire families. Indeed, my great friend and 'father' (fathers, mothers, brothers and sisters are the honorary titles given to very close friends in Central Asia), Professor Saim Balmukhanov (about whom I will say more later), was born in a yurt near Semipalatinsk, so this form of dwelling was still commonplace right up until Soviet times, when the nomads were dispersed and sent to live in collective farms.

Around 2000 BC, villages, towns and even small city states began to be formed in the more habitable and humid areas of Central Asia. The people who occupied these settlements had abandoned the nomadic way of life of their cousins who still roamed across the arid plains in favour of a more sedentary lifestyle, and practised irrigated crop production. There is evidence of early Bronze Age tribes such as the Bactria-Margiana and the Andronovo, who inhabited parts of Western Siberia and Kazakhstan and who invented the spoke-wheeled chariot. Gradually, these settlements grew into what became, by the first century BC, the Sogdian city states of the Ferghana Valley, home to the rich traders of the Silk Road.

The nomads grew to depend on trade with these city-dwelling cousins for goods which it was impossible for them to obtain otherwise. Where they had nothing of value to barter with the wealthy city traders, they would mount an assault and plunder and steal whatever they wanted.

Slowly but surely a diverse variety of tribes began to expand and populate the Central Asian steppe, including Turks, Huns, Persians, Scythians, Mongols and Indo-Europeans. All of the tribes had their own linguistic and ethnic variations, although it was clear that a broadly similar Central Asian culture was beginning to emerge.

In the second and first millennia BC, some large, powerful nations like the Macedonians and the Persians made repeated attempts to conquer the nomadic tribes of Central Asia, but their efforts met with limited success. Faced by big and powerful invading armies, the nomads simply melted away into the familiar territory of the vast Central Asian steppe, occasionally mounting savage guerrilla raids and taking a heavy toll on their unwelcome guests. With only their families and their livestock to defend, the nomadic tribes were a difficult people to conquer. It is a sobering thought that several millennia later, major military powers are still learning the hard way that bands of insurgent guerrillas can be a more dangerous foe than several divisions of heavily armed troops.

It wasn't until the time of Alexander the Great that the Hellenistic Greeks spread their influence as far as modern Tajikistan. Following Alexander's death in 323 BC and a series of bloody conflicts, India and China began to exert more control in Central Asia, sharing power with Greece and extending their base across a huge territory stretching from the Punjab to Afghanistan. It was here that the Silk Road began to expand and was able to forge powerful links between Europe and China.

A rich diversity of religious and cultural traditions could be found at this time in Central Asia. While Buddhism dominated, it was largely confined to the east. Zoroastrianism was the common religion in Persia, while Nestorian Christianity and other minor faiths were also in evidence.

Around the sixth century AD, the Turkic tribes of Central Asia began to expand their influence, taking control of vast areas of Asia and China. One of the dominant tribes at the time was the Uyghurs, the ancestors of the Turkic people of the same name who live in the Xinjiang autonomous region of China to this day.

It wasn't until the eighth century, and the steady encroachment of the Arab empire, that Islam began to percolate through the region. It soon became the sole faith of most of the population, though Buddhism remained strong in the east.

The Arab takeover also led to the expulsion of the Chinese from Central Asia.

Around this time, with the further development of the warhorse and advances in weaponry and horsemanship, the prowess of the nomadic armies began to be recognised. Over time, as new technologies were introduced, the nomadic horsemen grew in power and manoeuvrability, gradually evolving into one of the most powerful military forces the world had ever seen. Mounted archers who could ride a sturdy horse at speed, using only their legs for guidance, freeing up their hands to fire a bow or wield a sword or spear, became a fearsome adversary. Their skill and strength enabled them to travel vast distances, often covering up to 40 miles a day and continuing in this way for weeks on end, crushing everything that stood in their way.

These devastating armies ruled by powerful kings or khans quickly took control of Central Asia. The most renowned of these warriors was Genghis Khan (1162–1227), who united the tribes of Mongolia and eventually conquered large tracts of China, Russia and the Middle East, as well as controlling all of Central Asia. Unbelievably, by the time he died, Genghis Khan's empire covered one quarter of the landmass of the entire planet. It was this brutal, authoritarian leader who really set the historic precedent for the long line of similar harsh despots in this area right up to Stalin and beyond.

Genghis Khan's success was mostly due to his new and ultra-mobile tactics, which terrorised entire populations and vanquished all who dared to stand in his way. He organised his Mongol soldiers into units called Arav, comprising ten soldiers; Zuut, comprising a hundred men; Minghan comprising a thousand; and Tumen comprising ten thousand. Each of these units had a leader who was directly answerable to the next highest leader and so on, and it was mandatory for all males from age 14 to 60 to serve in Khan's forces.

These military units were made up of light cavalry, primarily composed of archers with little more than leather helmets for protection, and heavy cavalry, armed with swords and lances and heavy leather breastplates and helmets. Each cavalryman would lead a spare horse, which carried all his necessary supplies and equipment. This provided Genghis Khan with one of the most highly mobile and successful military forces the world had ever seen. Indeed their success could be measured by the fact that from his origins in Mongolia, where the population was less than 1 million, he was ultimately responsible for the deaths of an estimated 50 million people, or one-third of the inhabitants of every country he invaded. It wasn't until the reign of Stalin, more than seven centuries later, that this appalling record would be beaten.

Like Stalin, Genghis Khan employed an extensive network of spies and scouts who reported enemy strengths and locations. When reconnaissance detected a weakness, he massed his force of as many as 250,000 men and attacked, with the heavy cavalry leading the way and the archers supporting from the rear. Using these blitzkrieg tactics, Genghis Khan never lost a battle. These campaigns were often accompanied by wholesale massacres of the civilian populations. His army would sweep aside small, undefended towns and villages as they progressed, sometimes covering over 100 miles per day. The refugees from these towns would be driven towards the larger, more heavily defended cities, placing great stress on their supplies of food and water.

Next, Genghis Khan would dam or divert rivers, to deny water to the cities he was about to attack. He would then surround the city walls with massive catapults and siege engines, built on-site by captured prisoners. When he set up camp outside a besieged city, the Mongol leader would issue the order to surrender from a white tent. If the city surrendered, everyone, except the overall leaders, would be spared. On the second day of the siege, he would issue his order to surrender from a red tent. This meant that if the city complied, all of the men would be killed, but the women and children would be spared. On the third day, he would move to a black tent. After this, there would be no quarter given to anyone. Everyone would be killed and the city would be razed to the ground as a lesson to others.

Prior to his death, Genghis Khan followed tradition and nominated his successor – Ögedei Khan – and split his territory into khanates which were inherited by children and grand-children. When he died in 1227, after conquering the Western Xia, Genghis Khan was buried in an unmarked grave at an unknown location in Mongolia. Supposedly, the soldiers who organised the burial were systematically massacred in order to prevent the whereabouts of his grave being known.

Genghis Khan's descendants continued to increase the al-ready vast Mongol empire by conquering or subordinating large swathes of Eurasia, often by slaughtering much of the local population of the conquered states. At its largest, the Mongol empire included modern-day China, Korea, the Caucasus, Central Asia and substantial portions of modern Eastern Europe, Russia and the Middle East.

But it was not just his military accomplishments and fierce brutality that gained Genghis Khan worldwide notoriety; he also advanced the Mongol empire in other ways. He decreed the adoption of the Uyghur script as the Mongol universal writing system. He also promoted religious tolerance and created a unified empire from the nomadic tribes of north-east Asia. Present-day Mongolians regard him highly as the found-ing father of Mongolia.

After Genghis Khan died most of Central Asia continued to be dominated by the successor Chagatai Khanate. This state proved to be short lived, and in 1369 Timur, the next re-nowned and venerated leader from Central Asia, conquered most of the region. Timur (1336–1405), known in the west as Tamerlane or 'Timur the lame', was of Turkic origins and emulated the successful military traditions of his Mongol predecessors.

Timur had as fearsome a reputation as Genghis Khan, burning ancient cities to the ground after massacring their entire populations. By bizarre contrast, he was also a great patron of the arts, assembling a legendary collection of books and art, and sponsoring some of the most lavish and fabulous architecture of the period. Visitors to modern-day Samarkand in Uzbekistan can only gaze in awe at the beauty of these buildings regarded as some of the world's most historic master-pieces.

Timur's military conquests took him from Central Asia across Persia to Russia, and even to the modern-day Baltic states, where he conquered Lithuania. At one stage, while he was busy conquering cities in Russia, the Persians revolted. A vengeful Timur quickly returned to restore order, levelling entire cities and using the skulls of the massacred citizens to build gruesome pyramids.

In 1398, Timur's army of 90,000 troops crossed the Indus River and attacked India. They left a scene of carnage and devastation in their wake, wrecking the city of Delhi and seizing tonnes of treasure and 90 war elephants, which he proudly took back to his capital in Samarkand. The following year he turned his attention to Azerbaijan and Syria, then went on to Baghdad, which he sacked in 1401, killing more than 20,000 of its inhabitants, a feat that even Saddam Hussein was never able to emulate. Following this savage conquest, he went on to capture Turkey and Egypt. Around this time, the rulers of Spain, France and other great European nations began to tremble at the thought of Timur, who was now poised on their doorstep. They sent emissaries laden with gifts to try to pacify him and keep him at bay.

But Timur had other ideas. He decided in 1404 that he would conquer Ming China. His vast army set out in December, braving the blizzards and ice-storms of an unusually cold winter. Men and horses died of exposure, and the 68-year-old Timur fell ill. He died in February 1405 at Otrar, in Kazakhstan. Timur's large empire collapsed soon after his death.

The Gur-e Amir is the fabulous mausoleum erected to commemorate the Asian conqueror Timur in Samarkand, in modern-day Uzbekistan. It occupies an important place in the history of Islamic architecture as the precursor and model for later great Mughal architecture tombs, including Humayun's tomb in Delhi and the Taj Mahal in Agra, built by Timur's descendants, the ruling Mughal dynasty of North India.

The lifestyle that had existed largely unchanged since 500 BC began to disappear after 1500. An important change in the world economy in the fourteenth and fifteenth centuries was brought about by the development of maritime technology.

Ocean trade routes were pioneered by the Europeans, who were cut off from the Silk Road by the Muslim states that controlled its western termini. The trade between East Asia, India, Europe and the Middle East began to move over the seas and not through Central Asia. The disunity of the region after the end of the Mongol empire also made trade and travel far more difficult, and the Silk Road went into steep decline.

An even more important development was the introduction of gunpowder-based weapons. The gunpowder revolution and the advent of guns and artillery allowed the people of the towns and cities to fight off, and often defeat, the nomadic horsemen of the steppe. Construction of these weapons required the infrastructure and economies of large societies and was thus impractical for nomadic peoples. The domain of the nomads began to shrink as, beginning in the fifteenth century, the settled powers gradually began to conquer Central Asia.

Around the early part of the nineteenth century the Russians began to expand south into Central Asia, gradually taking over large tracts of the steppe and handing the land to Russian farmers. Thus began the process of major population movements to and from Central Asia which became a central hallmark of the Stalinist terror and to this day has left a diverse mix of cultures and religions living in the region.

The only real opposition faced by the Russians at this time was from the British, who feared that Russia was becoming too powerful and getting too close to the borders of the British Empire in India. The British and Russians began an energetic process of competition, which came to be known as The Great Game, as they both tried to advance their interests in the region. Nevertheless, the Russians managed to cross the Oxus River (the Amu Darya today), although their influence stopped short of Afghanistan, which remained as an independent buffer state separating the two rival empires.

Due to the American Civil War, the price of cotton rose dramatically in the 1860s, as it became an increasingly important commodity. However, its cultivation then was on a much lower scale than during the Soviet period. The cotton industry in the Soviet Union led to the construction of the

Transcaspian railway from Krasnovodsk to Samarkand and Tashkent, and the Trans-Aral Railway from Orenburg to Tashkent. In the long term, the development of a cotton monoculture would render areas such as Turkestan dependent on food imports from Western Siberia, and the Turkestan-Siberia Railway was already planned when the First World War broke out.

Russian rule still remained distant from vast areas of the region, and even then it only bothered about the small minority of Russian inhabitants. Local Muslims were not regarded as full Russian citizens. Although they did not receive many of the privileges given to Russian citizens, they were exempt from conscription.

During the First World War, the Russians removed this exemption, which prompted the Central Asian Revolt of 1916. When the Russian Revolution began in 1917, a provisional Government of Jadid Reformers, also known as the Turkestan Muslim Council, met in Kokand, a city in Fergana Province in Eastern Uzbekistan, and declared Turkestan's autonomy. They were swiftly defeated by Soviet forces, but guerrillas known as Basmachi continued to fight Communism until 1924. Mongolia was also caught in the Russian Revolution and, though it never became a Soviet republic, it became a Communist People's Republic in 1924.

Although there was an initial threat of a Red Army invasion of Chinese Turkestan, the region's governor agreed to cooperate with the Soviets. The creation of the Republic of China in 1911 and the ensuing unrest severely impacted Central Asia, and opened the region to the threat from both Islamic separatists and communists. Eventually the region became largely independent under the control of the provincial governor. Rather than invade, the Soviet Union established a network of consulates in the region and sent aid and technical advisors.

After being conquered by Bolshevik forces, administrative reorganisation occurred throughout Soviet Central Asia. This led to the creation of the Turkestan Autonomous Soviet Socialist Republic and the Soviet Socialist Republics of Bukhara and Khiva. In 1920, the Kirghiz Autonomous Soviet

Socialist Republic, covering modern Kazakhstan, was set up. It was renamed the Kazakh Autonomous SSR in 1925.

In 1924, the Soviets created the Uzbek and the Turkmen SSRs. This was followed in 1929 with the splitting of the Uzbek SSR into two separate units, one of which became the Tajik SSR, leading to problems which have continued until this day. The Kyrgyz Autonomous Oblast became an SSR in 1936.

This administrative reorganisation effectively created the basis for today's five Central Asian republics. Ethnicity was largely ignored when the borders were drawn but the Soviets did see the Turkish and the Islamic influences as threats, and divided the two accordingly. Under Soviet rule, the local languages and cultures were systematised and codified and their differences clearly demarcated and encouraged. New Cyrillic writing systems were introduced to break links with Turkey and Iran.

Under Stalin at least a million people died, mostly in the Kazakh Soviet Socialist Republic, during the period of forced collectivisation. Islam, as well as other religions, was also attacked. In the Second World War several million refugees and hundreds of factories were moved to the relative security of Central Asia, and the region permanently became an important part of the Soviet industrial complex. Several important military facilities were also located in the region, including the nuclear-testing facilities of the Polygon and the Baikonur Cosmodrome. The Virgin Lands Campaign, starting in 1954, was a massive Soviet agricultural resettlement programme that brought more than 300,000 people, mostly from the Ukraine, to the northern Kazakh Soviet Socialist Republic and the Altai region. This created a major change in the ethnicity of the region.

From 1988 to 1992, a free press and multiparty system developed in the Central Asian republics as perestroika pressured the local Communist parties to open up. However, the so-called Central Asian Spring was very short-lived, as soon after independence former Communist Party officials recast themselves as local strongmen. Independence largely resulted from the efforts of the small groups of nationalistic, mostly local intellectuals, and because Moscow had little interest in retaining this highly expensive region.

Increasingly, other powers have begun to involve themselves in Central Asia. Soon after the Central Asian states won their independence Turkey began to look east, and a number of organisations started building links between the western and eastern Turks. Iran, which for millennia had close links with the region, has also been working to build ties, and the Central Asian states now have good relations with the Islamic Republic.

Saudi Arabia has had an important influence on Central Asia, having largely funded the area's Islamic revival. Soon after the break-up of the USSR, Saudi money paid for massive shipments of Korans to the region, as well as the repair and construction of many mosques. In Tajikistan alone an estimated 500 mosques per year have been erected with Saudi money. The formerly atheistic Communist Party leaders have mostly converted to Islam, and Islamist groups have been created in many of the new states. Each of the new republics have remained largely secular, and all five states enjoy good relations with Israel. Central Asia is in fact still home to a large Jewish population which has cultivated important trade and business links with Israel and the Jewish Diaspora in other countries.

It is a sad irony that the kind, considerate and hospitable people of Central Asia were targeted by Stalin for some of the worst environmental atrocities the world has ever known.

CHAPTER TWO

The Polygon

The voice from the loudspeaker perched on a wall in the village of Kainar boomed out: 'Three seconds to detonation, two seconds, one second . . .'

'Explosion!'

'The shock waves carry frightening destructive forces!'

In 1953, Zhakiya Akhmetov was an eyewitness to the nuclear explosions. 'There was an enormous bang. I thought the earth was going to swallow us. Then we saw a very bright white light. It was as white as milk. So, I said to the others: "Look how white it is!"'

A year later the pensioner lost his eyesight completely. Now he says the sight only returns to him at night when he remembers the explosions in his dreams. Zhakiya served in the army and his unit was located at the test site, where they dug trenches and erected military buildings. After the obligatory service period, he retired from the army, but had a nervous breakdown. To this day, he speaks of nothing else but the Polygon, the name the Soviets gave to their vast, top-secret nuclear test site near the city of Semipalatinsk. His favourite story is about how wild ducks, blinded by the explosions, tried to fly but only collided with each other in the air.

From 1949 until 1990, the Soviet Union used the Semipalatinsk region of east Kazakhstan, near the border with Siberia, as a nuclear test site. Hidden from the world, this site the size of Wales was subjected to 607 nuclear explosions, including 26 above-ground tests, 124 atmospheric tests and 457 underground. Cynically, the military scientists would wait until the wind was blowing in the direction of the remote Kazakh villages before detonating their nuclear devices. KGB doctors would then closely study the effects of nuclear radiation on their own population, using them as human guinea pigs.

The Polygon is unique in global terms because it is the only former nuclear test site that can be readily visited by anyone interested. It is too vast an area to police. In some of the worst affected villages in the remote steppe, such as Znamenka, which lies about two hours' drive west of Semey (the new name for Semipalatinsk, the main city in the Polygon), many of the inhabitants are ill as a direct consequence of the bombs. In these villages it is possible to meet eyewitnesses, elderly men and women who explain graphically what it was like to witness the first atomic explosions and to see mushroom clouds rising a few miles away. They explain how they were ordered to stack bedding and furniture against their doors and windows to protect them from the shock waves, then to stand outside, away from any buildings, to watch the explosions. Similar accounts can be heard from residents of other villages affected by the tests such as Kainar, Sarzhal and Karaul, and even further afield in Ost Kamenogorsk and Urdzhar.

When the Soviet Union finally collapsed in December 1991, the departing battalions of troops and secret police who had guarded the Polygon left a legacy of devastation and sickness. The 1.5 million population of the Polygon had been subjected to the equivalent of 20,000 Hiroshima bombs. Seepage from the underground tests has polluted watercourses and streams. Farmland has been heavily irradiated. Radioactive contamination has entered the food chain.

Saim Balmukhanov

Professor Saim Balmukhanov is one of the leading academics in Kazakhstan. Now in his eighties, he recounts with clarity the impacts of the nuclear tests. 'The first indication we had that nuclear bombs were being detonated was in 1957,' he says. 'A doctor friend of mine from Semipalatinsk said that he had noticed strange discoloured lesions on the skin of some of his patients. He knew I had been involved in investigations following Hiroshima and Nagasaki and asked me to have a look. I confirmed that these were radiation burns.'

In one of my first meetings with Professor Saim Balmukhanov in mid-2000, he slammed his fist emphatically on the table. 'People in the West cannot begin to understand what we suffered under the Soviets. One and a half million people in Kazakhstan were exposed to high radiation doses during the Soviet nuclear-testing programme. But when I reported my findings to Moscow they denied it. For more than 40 years they claimed that the high incidence of cancers and babies born with genetic deformities were hereditary diseases caused by the poor Kazakh diet and an overindulgence in vodka.'

Until his retirement Professor Balmukhanov worked in the Kazakh National Academy of Science in Almaty, where he was first appointed professor and head of department in 1946 at the age of only 24. He was head of the Department of Biology and Medicine until 2005. His eyes sparkle as he speaks. He has made it his life's work to uncover the horrible legacy bequeathed to his people by the Soviet empire. Twice, this much-decorated war hero was arrested and stripped of his Communist Party membership when his enquiries got too close to the truth.

'Of course there had been rumours of explosions and strange mushroom clouds and village houses being swept away in the aftershock of the blasts,' he says. 'But people were afraid to speak out. The whole area around Semipalatinsk was closed and strictly controlled by the military. Villagers were told that they should be proud to be part of the great technological advances of the Soviet Union.

'There were sudden deaths and miscarriages,' the professor says, 'but each time we challenged Moscow they lied to us. When we checked the health of villagers within the Polygon – the 18,500 sq km territory of the core test site – against that of villagers from outside, we discovered there were four times fewer diseases outside the Polygon. Finally, in 1958, the Soviet military authorities had to admit responsibility.'

In one of our early meetings in Almaty, Professor Balmukhanov spread several large ledgers on the table of his study.

Each was prominently marked 'Top Secret' in Russian. 'These are the Soviet records of the human impact caused by their nuclear tests. The KGB right from the outset carefully recorded every piece of medical evidence. But all of it was highly classified and kept locked in a Moscow vault. Information, which could have helped us to treat patients and save lives, was withheld for 40 years. Only when the Soviet empire collapsed in 1992 did we finally gain access to this material. Even now, the Russians are still holding on to a lot of information,' the professor said.

My first meeting with Professor Balmukhanov took place in August 2000, when he described to me the plight of the people of Semipalatinsk in East Kazakhstan. He told me of the 607 nuclear bombs that had been detonated there, and how the Soviets had treated the local Kazakh population as human guinea pigs. I was both fascinated and horrified.

This was to be the first of many encounters with Saim Balmukhanov. Indeed, it is a tradition in Kazakhstan that when people become close friends they form a 'family'. Professor Balmukhanov now regards me as his 'son' and I look up to him as my wise and erudite 'father.' I visit my 'father' and his family every time I go to Kazakhstan. This first meeting with him came about almost by chance. It had been brokered at an earlier encounter in Brussels, shortly after my election to the European Parliament, in 1999.

Late in the afternoon during one hot, humid September in 1999, I was telephoned in my office in the European Parliament by a friend and former Green MEP from Germany, Frank Schwalba-Hoth. Frank asked if I had a few moments to spare, as he wanted me to meet a doctor from Kazakhstan. I protested that I was too busy, but when Frank persisted, I finally caved in and agreed, insisting that the meeting should last no more than 15 minutes. It was to be 15 minutes that changed my life.

Frank introduced me to Dr Kamila Magzieva, a striking woman with black hair and typical Mongoloid Kazakh features of light brown skin, high cheekbones and glossy black hair. As she described to me the plight of the people of Semipalatinsk and told me of the hundreds of nuclear bombs that had been detonated there and how the Soviets had treated the local Kazakh population as human guinea pigs, I was stunned.

I told Kamila that everyone in the West had, of course, heard of the enormous tragedy of Chernobyl, partly because the fallout had affected the West. But Chernobyl was a single nuclear explosion – an accident. I had never heard of the 607 nuclear explosions in Semipalatinsk. This was almost unbelievable. I asked Kamila if I could come to Kazakhstan and see the evidence for myself. She was overjoyed. She told me later that her meeting with me was the ninth she had held that day, which was 9 September 1999. Perhaps there was some sort of karma at work on 9-9-99. But of her nine meetings in the European Parliament, I was the only MEP who agreed to follow up her request to help her beleaguered people.

So it was in August 2000 that I found myself undertaking the first of many visits to Kazakhstan with Elena Kachkova, my trusted Russian interpreter and adviser. My visits would involve me in a campaign to draw attention to these victims of the Cold War and their appalling suffering and would welcome me to a new Kazakh 'family', with Kamila as my 'sister' and Professor Saim Balmukhanov as my 'father'.

During that first meeting Saim Balmukhanov told me that the people of the Polygon were suffering from a 'genetic multiplier effect'. When a man and a woman who have both been affected by radiation have a child, the genetic malformation in the baby is multiplied. He says that sometimes the damage may skip two generations, but then it will return with a vengeance. Many ill and severely deformed babies have been born in the past 20 years. He expects the next wave to appear around 2020, and anticipates that it may take at least until 2080 before the genetic impact begins to wear out, although no one knows for sure.

What we do know, he says, is that many people in the Polygon are ill. Cancers run at five times the national average. Birth defects are three times the national average. Babies and farm animals are born with terrible deformities. Children are mentally retarded and Down's syndrome is common. Virtually all children suffer from anaemia. Many of the young men are impotent. Many young women are afraid to become pregnant in case they give birth to defective babies. Psychological disorders are rife. Suicides are widespread. Average life expectancy is only 52 years. This is a man-made environmental problem of global significance which may take many decades to resolve.

The nuclear tests, which lasted for over a generation, had many detrimental effects, direct and indirect, on the health of the population of these regions. The direct effects are on people who were exposed to radiation during atmospheric and surface tests, those who lived on the contaminated land during and after the tests, and those born to them. These include cancers and mutations, as well as a variety of blood diseases and skin disorders.

Secrecy concerning the site and the impact of the testing on the population make it difficult to calculate the total population affected, but the number of those exposed to radiation was estimated at 1.7 million. Of these, 67,000 people were identified as having been exposed to the heaviest radiation. Out of the 67,000, there were 27,000 survivors, along with 39,600 descendants of their second generation and 28,900 of the third, totalling 103,500 people.

The indirect effects, on the other hand, are mainly the result of the closure of the test site and the disintegration of the Soviet Union. The economic stagnation and poverty that resulted from the closing down of the site contributed to a general state of malnutrition among the population and the weakening of their immune systems. As a result, for a time, the health situation in the Semipalatinsk Oblast (region) was the worst in the country. In 1999 the mortality rate was 11 for every 1,000 people, compared to 10.4 at the national level. The infant mortality rate was 28 for every 1,000 births, as opposed to a national average of 25.3. As for maternal mortality, it was around 286 per 100,000 live births in the region in comparison with 64 at the national level.

Birth defects are common, and actually increased for a time in the early nineties from 81 for every 100,000 people to 104.3, due to the genetic multiplier impact of having two parents both suffering from genetic malformation, arising from the exposure of their own parents to radiation. A variety of illnesses, including cancer, are rampant in the Polygon. The death rate from malignant tumours has constantly been higher than the national average. Endocrine, cardiovascular and infectious blood diseases, as well as a great weakening of the immune system, are commonplace. Over 80 per cent of the population in some areas is anaemic.

The incidence of mental retardation and psychiatric illnesses is also greater in the Polygon than elsewhere in the country. In 1996, the rate of psychiatric disorders was 21.5 per cent higher than the national average. The mental disorders tend to be of two types, either due to cognitive impairment or to psycho-traumatic factors. Nevertheless, anxious-phobic states and depression are prevalent, resulting in a greater suicide rate in the region. This has been particularly prevalent amongst children and teenagers.

Professor Balmukhanov has a rich source of stories about the people of the Polygon. During one of our many meetings he explained how once, during the Soviet era, he had secretly penetrated the security cordon around the Polygon so that he and some other doctors could investigate reports of the strange illnesses affecting the local villagers who lived in the zone. He set up camp next to the village of Kainar, in an abandoned overgrown orchard. 'We slept in decommissioned army tents, 20 people in each tent, and cooked our simple meals on the bank of a nearby stream,' he said.

'One evening, we were enjoying a well-deserved rest after a long and stressful working day. Suddenly we see a young man approaching us on an unsaddled galloping horse. He does not stop but shouts as he passes by, "Satai-Ata is asking for Dr Balmukhanov, please hurry, it's urgent!" Satai was the chief horseman in the small settlement of 30 to 40 people, and in charge of a herd of horses. The people are all closely related and obeyed his orders without question. The Kazakhs regard shepherding horses as an honorary job only entrusted to the most experienced people. Satai was a hereditary horseman; his grandfather was once in charge of horses belonging to a rich local landowner.'

He continued: 'By the far wall of a large room sat Zhamilya – The Beauty – the youngest daughter-in-law and the pride of the household. But her face was dark, and her eyes half-closed. When I said hello, she just nodded her head slightly in acknowledgement and did not utter a word. I sat down next to Satai in silence and waited. This is the custom; I was waiting to be spoken to.'

Professor Balmukhanov explained that Satai offered his apologies for having interrupted his important business. He then insisted that despite her appearance, his daughter Zhamilya did not need medical help. Nevertheless, Satai invited the professor and some of his medical colleagues to attend the ceremony of appealing to Aruakh (the Spirits of the Dead Ancestors) on the following Friday.

Professor Balmukhanov went on: 'After traditional tea-drinking the neighbours left, and Satai began to tell me about his sorrows. He had four sons. They grew up to be strong and handsome men. Now all are married. But he does not have grandsons to carry on the family bloodline. The only surviving grandson from his eldest son is mentally retarded. The other grandson from the same father could have had the makings of a true horseman. He had served in the army and was planning to wed a girl from Kainar.'

But this was not to happen, explained Satai. He told the professor that his 'beautiful son' had taken his own life last spring, a tear running down his wrinkled cheek while he spoke. 'And now Zhamilya,' he continued. 'She has been married for five years and has had four miscarriages. During her last pregnancy we all tried to help her with the chores so she could have more rest. Everybody got so excited as her pregnancy progressed, we were getting ready to celebrate the birth. But Aruakh had turned away from us and did not protect us from misfortune. The baby was born deformed, his legs fused together, with no toes. The old women confessed to me that the crippled baby was suffocated and buried far from the village.'

'And today,' Satai told the professor, 'while discussing husbandry over their tea, the men heard a noise from the front room, as if someone has tripped over a sack of grain. Then they heard the women scream "Oibai, Oibai, Ol'di, Ol'di!" ("She has died!"). They ran out to find Zhamilya collapsed on the floor with a noose around her neck. They quickly untied the knot and the young woman started to breathe. The rope must have rotted, or she did not attach it properly, but this suicide was not meant to be. Baibishe [Satai's wife] told us that Zhamilya was crying all week, complaining of her ill fortune. "I have brought a curse on your family," the poor girl used to say. "I wanted healthy strong children, like Satai." Her mother-in-law tried to comfort her, but in vain. Zhamilya decided to help her family by ending her own life.'

It turned out that giving birth to deformed children was not uncommon in this little settlement, Professor Balmukhanov explained. The old women could name seven or eight cases when crippled babies died soon after birth, their deaths probably being assisted by their relatives. There were only eight households in the settlement, up to 40 people, and Satai was the chief. That is why he decided to invite a mullah from a distant village, about 80 kilometres away. This mullah, unlike all the others, had retained the ability to ward off evil spirits. Satai was convinced that his small tribe had somehow managed to infuriate the spirits of their ancestors, who had then stopped protecting the villagers from evil.

'The exorcism ceremony was unknown to me,' said Satai. 'We had to invite Aruakh back with the sacrificial slaughter of a fat mare. The mullah then went around each house reading a great deal of prayers from the Koran. After the ceremony everybody had a celebratory feast.

'Aruakh is a collective name for the Spirits of Ancestors, who never leave us and help us in every good deed that we undertake,' Satai explained. 'When going on a long journey, the nomadic Kazakhs ask Aruakh to accompany them. Warriors call out Aruakh's name in battle, asking for help to destroy the enemy. The belief in Aruakh comes to Kazakhs from the Tengrian religion.'

'At that time I could not explain to Satai that the spirits were not to blame, but nuclear explosions,' Professor Balmukhanov said sadly. 'We did not have the knowledge then. Also, I could not predict that for many generations to come, while the bombs were being tested, it would be impossible for young women to produce healthy and beautiful children or strong and able horsemen like Satai.'

Balmukhanov continued: 'This was at a time when the Soviet Union was preparing itself for the Third World War. Nobody specified who the war was going to be with, but it was assumed the enemy would be the USA. A lot of lecturers descended on us from Moscow with the purpose of educating the authorities of the Republic of Kazakhstan on the consequences of nuclear attacks, the number of probable casualties, evacuation plans for the civilians, decontamination measures and so on.

'In 1952 I was appointed the chief radiologist for the Ministry of Health of the Republic of Kazakhstan,' said Balmukhanov. 'Prior to that, all heads of radiology departments in medical universities had completed a four-month crash course in medical radiology. I also attended the courses. During that brief period, we received detailed information about the consequences of atomic bombs dropped on Hiroshima and Nagasaki, about hundreds of thousands of Japanese people killed on the spot and about just as many suffering and dying from radiation sickness.

'We had to study the basics about radiological safety and hygiene measures and how to calculate the exposure of the population to nuclear radiation. After a few months working for the Ministry of Health, I was told about the existence of the Polygon – the Semipalatinsk nuclear test site. I voiced my concerns with the minister and was able to convince him that nuclear testing in the Polygon might be harmful to the well-being of the population in the entire region. The Polygon was top secret, and there was no way of obtaining any information about what was going on there. I was left with no choice but to conduct an evaluation of the health of the villagers in the region *in situ.*

'In those years, nobody without special clearance from

Moscow could enter the Polygon. I was issued with false documentation in the name of the head of the regional health department for Semipalatinsk. Together with two doctors from the regional hospital we set off to visit the Abai District. At the district centre in the village of Karaul, the doctors shared their concerns about a strange increase in neurological illnesses, unusual skin conditions and birth defects in the village of Sarzhal, located on the perimeter of the Polygon. They also told us that the village herds of cows and horses still grazed in the Polygon.

'We were slowly advancing towards our next destination along a dirt road. It was incredibly hot; the grass in the steppe was high and burnt by the sun; there was no sound around apart from the buzzing noise of thousands of insects. Suddenly we heard the revving of a car engine, and were overtaken by three UAZiks (Jeeps) and signalled to stop. The cars were packed with military officials, all of high rank. They demanded to see our clearance papers and enquired about the purpose of our trip. Our answers must have sounded suspicious to them, and the decision was taken to take us to Kurchatov, the top-secret city at the centre of the Polygon, to establish our true identity and intentions. The procedure itself was quite standard. Anybody caught trespassing was taken to Kurchatov and kept, from several days to several weeks, while the KGB checked them out. We tried to show them the paper from the Ministry of Health allowing us unrestricted movement in the area; we tried to talk about our professional and personal duties as doctors, but received the same answer: "You will be dealt with in Kurchatov."

'Unexpectedly, another car arrived. Amongst its passengers one man stood out – very energetic, of stocky build, with a bald patch amidst the mass of hair, dressed in civilian clothes. He looked like an Estonian, or a Pole. He quickly approached us and enquired about the purpose of our trip. For some unknown reason, I decided to trust this man; I wholeheartedly admitted the illegality of our being there. I also told him that I was a radiologist, had attended the recent training in Moscow and that I was concerned about the effects of nuclear tests in the region on the health of the locals.

'Having listened to my story, he moved back to the military officials and began a heated debate with them. The end result of this was that we were handed our confiscated documents back and allowed to continue our journey. Our saviour approached us once again and said, "A great misfortune has befallen your people. There will be many hardships and obstacles along the way to rectify this. I am involved in trying to resolve the issue. I will do my best to relieve the sufferings of the people, and I wish you every success in your endeavour." His words impacted deeply upon me. I remember them clearly to this day. He also shook our hands and introduced himself as "Sakharov".

'In the late autumn of 1953 the Regional Party Committee of Semipalatinsk gathered to discuss the results of a nuclear test conducted on 12 August 1953. This test involved the detonation of a gigantic 400 kilotons bomb, known as Joe-4 in the West, or RDS-6 in Moscow. There I learnt that our saviour was in fact a Hero of Socialist Labour, the highest civilian order, and the primary designer of the thermonuclear bomb. He was none other than the outstanding scientist Andrei Dmitrievich Sakharov. I also learnt that for the first time in the history of the Polygon the population of adjacent villages had been evacuated on Sakharov's insistence.'

Balmukhanov continued: 'Our second encounter took place in 1969. As a Fellow of the Kazakh Academy of Sciences and Director of the Oncology Institute, I was summoned to the office of the Almaty Committee of the CPSU (Communist Party of the Soviet Union). I was told my signature was required. When I asked what the signature was for, the Committee Secretary became irate: "Don't you read newspapers? Haven't you heard that Sakharov is an enemy of the people and is on his way to Siberia?"

'"What, exactly, is he accused of?" I enquired.

'There was a long pause, then the Party Secretary came up with the following words: "If you have doubts about the actions of the Party, you should hand in your Party ID and resign your membership immediately." I was given a page to sign; it was already signed by a lot of high-ranking doctors and

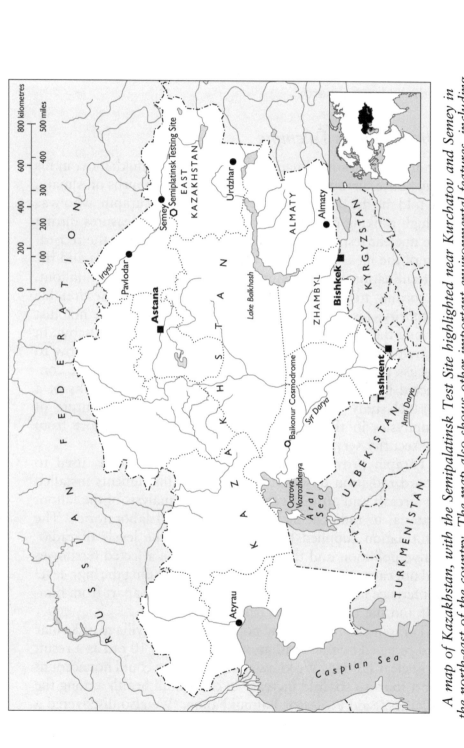

A map of Kazakhstan, with the Semipalatinsk Test Site highlighted near Kurchatov and Semey in the north-east of the country. The map also shows other important environmental features, including Lake Balkhash, the Aral Sea and Baikonur Cosmodrome.

scientists known to me. It contained a list of Sakharov's "crimes". I refused to sign it. That was my second encounter with Andrei Sakharov.'

The Colonel's Dilemma

At another meeting I had with Professor Balmukhanov, in his tiny house surrounded by apple trees in the suburbs of Almaty, he told me the tale of Colonel Sergei Lukich Turapin, who was a high-ranking officer responsible for safety measures during the nuclear tests. Amongst his responsibilities was the recording of the measurements of the force of the explosion and the determination of the boundaries of the radioactive fallout. According to Balmukhanov, Turapin had an unquestionable authority; his opinions were considered important, not just among the employees of the Polygon but also by Moscow. In March 1961 a top-secret conference was held in Moscow to discuss the findings of a three-year research programme conducted by the Kazakh Academy of Sciences, the results of a parallel study conducted by scientists from the Institute of Biophysics in the same remote villages and a report from the security services of the Polygon.

Turapin provided a detailed report on methods used to record radiation doses with the help of instruments installed on aircraft and ground vehicles, and on analysis of the fallout material in radiochemical and biophysical laboratories. The information supplied by Turapin about the levels of radioactive pollution and the boundaries of the affected territories did not cause any concern among those attending the high-level conference in Moscow, claims Balmukhanov, apart from himself and the other doctors from Kazakhstan.

Turapin claimed that the population of the village of Kainar had received minimal radiation doses of 7 to 10 rad as a result of nuclear fallout. 'We knew that such doses could not account for a four- to six-fold increase in cases of ill health among the villagers,' said Professor Balmukhanov. 'We also discovered a high content of radionuclides in the soil, plants and foodstuffs in the region. But the conference participants chose to believe

Turapin because of his high authority and his sophisticated measuring equipment. The only person who defended our report was academician Andrei Vladimirovich Lebedinski, Director of the Institute of Biophysics. He urged the participants not to dismiss the findings of the Kazakh scientists, which had been confirmed by the researchers at his institute. He said there was a mystery, a phenomenon that needed to be explained with further research.'

So a new term was born – 'the Kainar phenomenon' – which was to be used over many years. In reality, there was no 'phenomenon' at all, and the mystery was revealed by the same Colonel Turapin some years after. In 1989, the secretary of the regional committee of the Communist Party informed Professor Balmukhanov about a planned conference to be held in Semipalatinsk with the participation of staff from the Polygon, Kazakh scientists and representatives of the local population.

'I thought it would be a good idea to invite Colonel S.L. Turapin to take part in the conference,' said Balmukhanov. 'I knew that he was retired from the military service and had been given a small flat in the town of Zheltie Vodi on the Dnieper River. Sergei Lukich accepted my invitation with delight, and we then met again, 25 years after our first encounter, in Semipalatinsk. I learned from him that his retirement was forced upon him quite a bit earlier than usual after he began to express doubts about the accuracy of measuring techniques where the air samples were collected into special bags equipped with what was called the Goryainov filter.

'Such filters were produced in the USSR on a mass scale and were mainly used to reduce the expulsion of soot and smoke from boiler chimneys into the atmosphere. Turapin explained to me that a cloud formed as a result of a nuclear explosion had a great variety of differently sized particles. Particles of metal, rock and soil are all electrically charged. Having conducted a detailed comparison, Turapin arrived at the conclusion that the physical properties of ash, soot and smoke from the boilers were fundamentally different from those of the particles in the radioactive fallout.

'Depending on their thickness, the Goryainov filters could stop between 80 to 90 per cent of soot and ash, but the particles of radioactive dust had a far greater ability to penetrate the filter, with only 20 to 30 per cent being retained. Therefore, the doses of radiation received by the population in the region, in his opinion, must have been miscalculated, and should have been two to three times higher. He had based his PhD thesis on his findings and sent it to Moscow. But instead of a hearing, an order came for his immediate dismissal. Turapin then appealed to the Central Committee of the CPSU, but with no result. He could not even get his thesis back; it disappeared without a trace. On his demobilisation he was also made to sign conditions under which all his research work had to be stopped and all material had to be submitted to HQ.

'I advised Turapin to inform the local Party secretary about the topic of his report at the conference,' said Balmukhanov. 'He was scheduled to give his presentation after Lieutenant-General Ilienko. He was quite excited the night before, expecting at least a minor victory after all these years of silence. But "the bomb", which was Turapin's report, was never meant to explode. We were told that he had changed his mind and left Semipalatinsk during the night. Everybody was shocked by his actions. As we discovered afterwards, he had a midnight visit from KGB officials who reminded him of his signed conditions and that his address at the conference could be interpreted as exposure of state military secrets and would therefore amount to treason, with his immediate relocation to Siberia. What would you do? The poor Colonel packed his bag and left before dawn, without saying goodbye to me. He died the following year at home, in Zheltie Vodi, of a heart attack.'

In 1974, the United States and Soviet Union signed the Threshold Test-Ban Treaty, limiting the yield of underground nuclear tests to 150 kilotons. Two years later, in 1976, the two countries signed the Peaceful Nuclear Explosions Treaty. However, ratification of both treaties was delayed due to a lack of effective verification procedures. A comprehensive moratorium was only finally agreed at a summit meeting between Reagan and Gorbachev in 1990. In the intervening years, the Soviets had cynically continued to test atomic weapons, claiming that they were carrying out peaceful underground explosions in the Polygon to construct a lake, in order to supply fish to the local population.

Thus the so-called 'Atomic Lake' was born. This massive radioactive reservoir was blasted out of the low-lying mountain range which crosses the steppe in the region of Semipalatinsk. The Soviets even tried to introduce fish to the highly radioactive waters, encouraging local Kazakh villagers to catch and eat their deadly harvest. A group of elders in the village of Kainar told me how a Soviet general, determined to prove to the world that the waters of the Atomic Lake were safe, called a press conference, then stripping off his clothes, dived into the shimmering water and swam for several hundred metres. He died of cancer ten months later.

Now there is growing evidence that cracks and fissures in the geological strata of the Polygon have allowed plutonium, strontium and americium into the River Irtysh, which flows from China, through the Polygon and on through Siberia to the Kara Sea, and eventually into the Arctic Ocean. The Soviet nuclear legacy may yet cause a world catastrophe.

Kharitonchiks

Professor Balmukhanov once told me: 'You will not find the word Kharitonchiks in any dictionary. It is a word solely used by the staff of the Semipalatinsk nuclear test site in the Polygon. The word was coined by scientists working under the supervision of Yuli Khariton, who was in charge of the Soviet nuclear test programme for 45 years. He was based at Kurchatov.

'In 1989, the Ministry of Health of the USSR sent a medical research team to the Semipalatinsk Region. It was headed by Professor A.F. Tsyb, Director of the Radiology Centre in the city of Obninsk [known as the first Russian Science City]. For one month Professor Tsyb was ordered to conduct a detailed medical examination of several thousand residents in the Semipalatinsk region. They were also to measure levels of radioactivity in most of the settlements in the region, numbering a few dozen remote villages. The time period allocated for the research was clearly not long enough, and a lot of the activities of the research team were conducted in a hurry by overworked scientists toiling for 16 hours a day, affecting the objectivity of the findings.

'The authorities in the Polygon offered Professor Tsyb the facilities to measure radioactivity at any site he chose. The Soviet military provided him with a helicopter to visit almost every single site where a nuclear test had been carried out. Around the towers and tunnels, within the radius of one hundred metres, he found bits of rocks and minerals melted into shiny multicoloured pieces of glass with the force of the explosions. In one of the tunnels he was amazed with the diversity of shape and the rainbow colours of glassy stones. He and his team stuffed them eagerly into their pockets, like schoolboys collecting coloured marbles.

'As Professor Tsyb and his colleagues approached the helicopter to take them back to Kurchatov, the official from the test site who was allocated to their team, whom they suspected was a KGB officer, stopped them and said: "Now quickly take all these Kharitonchiks out of your pockets and throw them away. Don't make me frisk you important scientists, it would be embarrassing for all of us!"

'Professor Tsyb, who was known for his fiery temperament, replied: "What kind of rubbish are you talking about? What on earth are Kharitonchiks?" The official explained: "We call Kharitonchiks all those small pieces of rock, all the glassy objects thrown out of the tunnels by the shock waves of the nuclear explosions." So Professor Tsyb and his colleagues had to obey his order, and disposed of their Kharitonchiks with some regret. It is just as well that they did so; it probably saved their lives.'

The Institute of Nuclear Safety and Ecological Studies of the National Nuclear Centre of the Republic of Kazakhstan, to give it its full title, is situated in the city of Kurchatov, the remote city specially built by the Soviets to house their nuclear scientists and KGB minders. Its specialists view the Polygon as a natural scientific laboratory, stating that the experiments were carried out in the local environment, and we have no alternative but to accept it. Here they say that the problems associated with nuclear tests are the things of the distant past, and that the Polygon is not dangerous to the health of the people living around it any more.

The director of the institute is Sergei Lukashenko: 'Today, let's say, 99.9 per cent of the territory is completely safe, there is no danger of irradiation on it, one can drive in, have a look, go for a walk there and enjoy the steppe with its unique vegetation,' he says, rather disingenuously. But he admits that there are still some highly dangerous places in the Polygon.

One of the most dangerous spots is the Atomic Lake, formed in the hilly region of Degelen as a result of a thermonuclear explosion carried out in 1965. This was the first industrial explosion, allegedly designed to address the needs of agriculture in the region. It was planned that the waters from the lake could be used for irrigation purposes in the drought-affected areas of Kazakhstan, and also to supply drinking water for farm animals.

The test was conducted underground at a depth of 178 metres, its yield being 140 kilotons. Today Mr Lukashenko insists the water in the lake is not radioactive, although he warns that the situation around the lake is far from safe. Even now, almost half a century after the test, the level of background radiation at the test site is 500 micro-Roentgens, which is ten times higher than permissible levels. Also, the soil on the banks of the lake is radioactive.

According to Yuri Strilchuk, Head of the Department of Nuclear Research at the Kurchatov Institute: 'The soil is contaminated with radionuclides. They get into plants and grasses, and, of course, you can see pastures on your way to the atomic lake, only six or seven kilometres from the place, where sheep and cattle are allowed to graze. In principle, there we have a dangerous situation with internal exposure to radionuclides by meat and milk.'

Yuri Strilchuk confirms that there are various other ways for radiation contamination to affect the people of the Polygon.

'For example, at the Balapan [Atomic Lake] site, we have more than a hundred vandalised shafts where the charges for atomic bombs were placed to conduct underground tests. The shafts used to be filled with concrete, but the people dug them up, sawed off metal pipes and took them away to unknown destinations. This presents a real danger to the population and exposure to radionuclides, because radioactive material is being taken outside the Polygon territory and nobody knows where it is exactly.

Says Lukashenko: 'There are no such places on the Polygon which you can enter and immediately feel the effect from exposure to radiation. There are, however, cases of careless and unacceptable behaviour by the local population, resulting in dangerous levels of overexposure. This is why we try to educate the people and to point out places where any activity is permitted and those where some activities are restricted, as well as those where any kind of activity is strictly forbidden.

'There is talk amongst people here about a three-legged, two-headed calf. I myself studied in Leningrad, or St Petersburg, as it is called now, where I frequently visited the Kunst Kamera, a museum where Peter the First exhibited his collection of deformed people and animals. There also they have a two-headed calf, but that calf dates back to 1700, when no one had heard of radiation. Personally, I do not approve of such speculations.'

Blind Steppe Eagles

Saim Balmukhanov told me a story about an event that happened in the summer of 1952. 'I went to see the regional secretary of the CPSU Committee, Comrade Mukhamedzhan Suzhikov. Despite his high position, he, unlike many others, was more accessible. As far as I was concerned, he always agreed to see me without delay and I often came to him with my concerns regarding the Polygon.

'This time, I came to express my concerns about the practice of unilaterally signing the Act of the Successful Testing of

"Devices". The act required signatures from the regional secretary, district secretary and secretaries of the bordering districts and from us, representatives of the Ministry of Health of Kazakhstan.

'The Act was compiled by the specialists working in the Polygon, and contained information on the date and place of the explosion, the power of the "device" and the trace of the cloud formed after the explosion. What caused my concern was that this act contained a statement (which I had to sign) which implied that there would be no detrimental effect on the health of the people in several villages adjacent to the explosion in the Polygon. Therefore, in front of comrade Suzhikov, I stated my intention not to sign the act, as I did not inspect the people living in the area covered by the radioactive cloud fallout.

'Comrade Suzhikov fell silent for a long time, we even managed to drink a few cups of tea while he was pondering his response, and he did not come up with any! The next morning I was summoned to the Regional Committee [Obkom]. At the door I was met by the chief doctor of the regional hospital, who informed me that we had permission to visit the controlled zone – Sovkhoz Socialistik, a collective farm some 40km away from Semipalatinsk. We were even given permission to gather information about the state of health of the residents of Socialistik and to conduct arbitrary check-ups amongst them. The local district secretary of the CPSU Committee and the director of Sovkhoz were made aware of our visit – in those years there was a protocol about controlled movement of any kind in the area.

'Having presented the relevant papers at the Semipalatinsk checkpoint, we left the town in a horse-drawn carriage and hit the steppe. There was no road as such, and our ride was very bumpy. The steppe was flat and our route monotonous, along the road marked by telegraph poles, on top of which the steppe eagles proudly sat, with smaller birds like sparrows and swallows taking their place along the cables. The road was winding and when our carriage approached the poles, disturbed eagles lazily took off and reluctantly relocated to the nearest pole. By the end of the day we reached the village of Socialistik, the central village in the sovkhoz of the same name. The steppe eagles were still sitting on the poles here, but they looked different, with their heads buried in their chests and their feathers spread as if ruffled; because of this they seemed even bigger. But I was mostly amazed at how "docile" the eagles were; even when our noisy carriage approached the poles, they did not fly away.'

Balmukhanov explained what happened next. 'When we reached the local hospital, I spoke to the head doctor. "How did the villagers react to the testing of the 'device'?"

' "The same way as always," he replied. "The men drank themselves stupid throughout the week, so they did not notice anything. I was forced to admit several hysterical women to hospital, all with the same complaint of uncontrollable vomiting and nausea. Some other neurotic women were complaining about constant migraines. Well, we took their blood pressure – they all had hypertension."

' "Why were your men drinking?"

' "The vodka is free, so why not? A lieutenant at the nearest checkpoint supplies them with drink. Says it's his own ration . . . Who knows!"

' "Listen, I was very curious about those tame eagles who sat on the poles calmly when we passed them."

' "But they are blind!" the doctor replied, "and sick. The village boys knock them down with sticks and stones, or sometimes they just die and fall to earth. The silly birds watch the nuclear explosion and go completely blind within several minutes."

' "What are those silly, inquisitive birds guilty of? All they do is control the population of rodents which thrive in the steppe . . . Those blind steppe eagles fall down dead, the boys knock them down with stones, just to vacate their seat for new eagles, which, in turn, will be blinded by the next test of the "device".'

CHAPTER THREE

The Nuclear Legacy

Semipalatinsk

Getting to Semipalatinsk near the Siberian border in Kazakhstan is always an interesting experience. The two-hour flight from Almaty is in a YAK-40, operated by Kazakh Air. These are 40-seat workhorse jets which were left behind by the Soviets when they departed from Kazakhstan in 1992. They have no hold, so all bags and suitcases have to be carried on board and stacked in a sort of curtained-off shelf at the rear of the plane, which you enter by climbing up a steep gangway in the tail, like crawling up the backside of a crouching dragon. The first time I made this trip the plane was full, so one woman had to sit on the floor in the central aisle for the duration of the flight. On other occasions I have seen passengers ordered to stand for the entire trip and grip a special 'monkey bar' that runs along the ceiling of the cabin.

In August 2003, we were seated at the front of the plane on hard steel seats, with old loose cushions the only attempt at comfort. I was accompanied by Kamila Magzieva, my Kazakh 'sister', Elena Kachkova, my interpreter, and this time also by a great friend, the Hollywood actress, photographer and humanitarian Kimberley Joseph. I had met Kimberley on a flight from London to Melbourne, where she had been invited to open the Melbourne Races, being the celebrity star of *Gladiators* and other popular Australian TV shows. I spent the entire flight boring her with tales of Semipalatinsk and the Polygon, and to my great surprise, when we landed in Melbourne, she said, 'Next time you plan a trip to the Polygon, give me a ring and maybe I will come with you.' To my even greater surprise, she had taken time out following her major role in the UK television series *Cold Feet* to join me on this latest sortie into the depths of East Kazakhstan.

Back on the plane we were fanning ourselves frantically with magazines and newspapers, as the searing heat inside the aircraft rose to dangerous levels. Our pilot, dressed in a dirty T-shirt and jeans, appeared to be arguing with the driver of a fuel tanker parked beside the plane. When he squeezed past us to enter the cockpit, Elena asked him, in Russian, what was the problem. He explained that he'd tested the fuel in the tanker before filling up the plane and discovered it had been watered down! He'd ordered a new tanker to come and we'd have to wait until it arrived.

As we digested this rather unnerving news, a Kazakh pop singer, who was also a passenger in the plane, opened a huge bottle of vodka, took a deep slug, then started to pass it around the entire aircraft. Elena, never a good flier at the best of times, was particularly grateful for this small swig of 'Kazakh courage'.

On arrival in Semipalatinsk, we were met by the *akim* (mayor) and a group of local girls in traditional green and gold Kazakh costumes. The press were there in large numbers to record our arrival, and Kamila, Kimberley, Elena and I were presented with bouquets of flowers and showered with sweets and other gifts by the costumed girls. Akim Omarov whisked us off to a riverboat moored on the River Irtysh where a table had been heaped with horsemeat, a sheep's head, nuts, sweets and other local delicacies. Virtually all of the senior local Kazakh politicians, academics and doctors had been invited to the lunch. Endless toasts were made throughout the meal, and each time I drained my glass of vodka an attentive steward would refill it. After about fourteen such toasts, I was alarmed to be told by Omarov that we had to hurry because we were already ten minutes late for an international press conference!

Following the fairly incoherent press conference, we were taken to the main Semipalatinsk hospital by the chief medical officer, Marat Sandybaev. An elderly patient – Kizat Kuzembayev – stood proudly to attention as we entered his tiny ward in the cancer wing of the hospital. Medals were pinned to his dressing gown, indicating his status as an important war hero. He was 79 years old and suffering from terminal stomach cancer. In front of two other elderly cancer patients who shared his room, he explained how he had served with a reconnaissance unit in Danzig during the Second World War, receiving the Order of Glory, the Order of the Red Star and the Great Patriot's War medal in recognition of his bravery. These were the highest decorations for ordinary soldiers in the Soviet army.

But in 1953, he was one of forty-two healthy young men selected by the Soviet military regime to act as human guinea pigs. The small group was taken to the village of Karaul in the remote steppe of East Kazakhstan. Local villagers had been evacuated, and Mr Kuzembayev and his colleagues were ordered to leave the shelter of the village houses in which they were billeted to watch an atomic explosion from a nearby hill. They were only 30 miles from the test site.

Mr Kuzembayev recalled the nuclear blast in vivid detail. He saw the sky turn red as if a huge fire had engulfed the landscape from horizon to horizon. As the ground trembled beneath his feet and the hellish roar of the atomic explosion swamped Karaul, he watched the fiery sky turn black, then grey, with piercing white and red spirals of flame shooting upwards, while the writhing stalk of the monstrous mushroom cloud unfolded. Later, KGB officers told his group that they would now have 'no worries from the USA', as the Soviets had perfected their own atom bomb.

Mr Kuzembayev told us he felt fortunate to have lived to see his 80th year. He was the only surviving member of this group of nuclear guinea pigs. The other 41 had each died of cancer. We told Mr Kuzembayev that we intended to visit Karaul on Thursday and he immediately asked Dr Sandybaev for permission to leave hospital for two days in order to meet us there. Despite his obvious illness, permission was granted.

Later we visited the museum to victims of the nuclear tests. Housed in a small building near the children's hospital in Semipalatinsk, it is approached through a garden where, rather incongruously, all the old Soviet statues have been dumped. Visitors to the museum must pick their way gingerly past vast statues of Marx and Lenin, and even giant, heroic-looking sculptures of Stalin, arms outstretched as if to embrace the citizens of Kazakhstan.

It is a great irony that a few yards away, rows and rows of glass jars contain the gruesome exhibits that bear grim witness to Stalin's legacy. Here in the horrific collection of the museum are some of the victims of the nuclear tests. Deformed babies, born with their brains, intestines or spinal cords exposed. A boy born to a Soviet pilot and his wife, who were working on the nuclear tests, with a single eye in the centre of its forehead – a perfect Cyclops – a harrowing reminder of the enormity of the radiation damage he and his parents were subjected to over a long period of time. This poor, deformed baby was born alive and even survived for a few hours.

Next morning Akim Omarov had arranged for a convoy of cars to take us to the Andas-Altyn gold mine a few kilometres from Semipalatinsk. The Andas-Altyn Mining Company, a Scottish–Canadian–Kazakh operation in association with the local authority in Semipalatinsk, opened the mine in January 2000. By August that year they were employing 530 Kazakhs and paying good wages. They told us they had mined almost one tonne of gold in the past six months. Akim Omarov said he was keen to see further inward investment to his region, and he was particularly proud that the gold mine had opened because most businesses were scared to move to the Polygon due to the radiation risks.

However, he admitted the mine workers were still being exposed to dangerously high levels of radiation because the underground nuclear tests had contaminated the rock strata and penetrated the water courses. 'But at least these miners have a job,' he said. 'For tens of thousands of innocent, unemployed Kazakhs, the legacy of the Cold War is one of suffering and hardship. There is little inward investment. Few have the courage to build a business in a former nuclear test site.'

Later that evening, one of the most senior officials in the regional government of Semipalatinsk invited us to his dacha, or summer villa, overlooking the River Irtysh. Once again a table had been laid with great quantities of food, including an enormous cooked sturgeon and a large bowl of caviar. In a corner a quartet of musicians was quietly playing Kazakh music. After the usual vodka toasts and lavish praise for the 'important visitors from Europe', the senior official asked Elena, my interpreter, and me to step outside onto the veranda. He slid the glass door shut behind us and, speaking in Russian, enquired if I had enjoyed my visit to the Andas-Altyn gold mine. I confirmed that I had. He then startled me by asking if I would like to have shares in the mine.

'I cannot afford to buy shares in a gold mine,' I answered.

'I was not suggesting you buy the shares,' said the high-ranking official. 'I will give you the shares as a gift from me. The city of Semipalatinsk owns 51 per cent of the stock in the Andas-Altyn Mining Company and it will be my great privilege to give you some of this stock as a sign of my respect for you, my brother!'

'I am a politician,' I protested. 'How can I accept shares in a gold mine?'

'I too am a politician,' exploded the official. 'What is your problem? We cannot be expected to live on the poor salaries we are paid, so we have to look after ourselves.'

I turned to Elena and said, 'Please explain to him that I do not accept bribes.' I slid open the door and walked back inside the dacha. Elena slid the door shut behind me, but even over the noise of music from the quartet and chatter from the other guests, I could hear Elena shouting at the official in Russian. She told me later that she had given him a severe bollocking, yelling at him for daring to offer a bribe to a Western politician, and that he was both shocked and surprised at my reaction.

Elena is Russian. She is the daughter of two academics, who actually were both rocket scientists, and she was brought up in the closed city of Novosibirsk in Siberia. She served her compulsory military training in the Red Army, where she rose to the rank of captain, so she is a tough lady and takes no bullshit from anyone. She was certainly a match for this senior apparatchik, no matter how lofty his position.

But this was not our only encounter with this official. Later that week we were invited by the owners of the gold mine to a banquet in a local sauna. Saunas are an important way of life for Kazakhs, and they frequent them day and night throughout the year, even in the hottest summer months. They are more like the health clubs or gyms we know in Europe, with elaborate plunge pools, steam rooms, saunas and restaurants. The directors of the gold mine had hired an entire sauna on the outskirts of Semipalatinsk for the night and caterers had prepared the usual grand buffet of sturgeon, horsemeat and caviar.

The guests were a mixed company of businessmen, politicians, local professors and doctors, male and female, but all of us were told to strip off, sit naked in the roasting sauna until the sweat was pouring from us, then plunge into the deep and almost heart-stopping frozen plunge pool, before wrapping a towel around our bodies and heading to the table for food and vodka. This process was repeated endlessly for the next few hours. Late in the evening I was again accosted by the same high-ranking apparatchik. He summoned Elena to interpret.

'Do you like beautiful girls?' he asked.

'What do you mean?' I enquired. 'I am a happily married man.'

'I am also a happily married man,' yelled the exasperated official. 'But Kazakh men can be happily married and still enjoy the company of beautiful girls. I am going now to a special club where we can meet these girls and I am offering to take you with me.'

Once again I asked Elena to intervene as I beat a retreat to the sauna; I saw her standing stark naked in front of the giant and also naked official, shouting at him and waggling her finger under his nose, demanding an apology for daring to insult my integrity by making such a proposition. The disgruntled official shuffled off to the men's changing room and soon emerged fully dressed, to lead a party of his friends off to the girls' club.

Next morning, the chastened apparatchik apologised to me and said, 'I cannot seem to interest you in any of the gifts that I had wished to give you, so now I feel embarrassed to have offered you such things, which are normal for us here in Kazakhstan.'

So saying, he presented me with a lethal-looking hunting knife and a traditional Kazakh knight's embroidered gown and peaked hat, insisting that I try it on. He then had the press take photographs of me dressed up like Merlin the Wizard!

Chagan and Ground Zero

Driving out of Semipalatinsk over the cracked and potholed road, the crumbling ruins of the Cold War soon become apparent. The road to 'Ground Zero', where the nuclear weapons were detonated, stretches for hundreds of kilometres across the barren steppe. 'Ground Zero', our guide from the local city hall explained, using the English words, is the name local villagers have given to the place where in 1949 the first Soviet nuclear device was exploded on the Semipalatinsk Test Site. Thus began the nuclear history of the Soviet Union. The capacity of the bomb was 20 kilotons. Even today, more than six decades after the event, the level of radioactivity here exceeds by a thousand times what is considered to be the safe, normal background level. This is the most contaminated spot in the Polygon, the epicentre of nuclear testing.

During the Soviet era, massive security surrounded the Polygon. Whole cities were erected to house military and scientific personnel. Their names never appeared on any maps. Residents were forbidden to mention where they lived, even to a neighbour from a nearby village.

About 80km from Semipalatinsk is Chagan, a city built between 1947 and 1949 as a base for the Soviet army and air force, and now completely deserted and derelict. Street after street of broken tenements bear silent witness to the nuclear arms race. Weeds sprout from cracks between crazily rearing flagstones. A statue of Lenin tilts dangerously to the side, the nose broken off and the base scrawled with graffiti. In the crumbling foyer of the cinema, a huge reel spews strips of brittle film onto the cracked floor tiles. I picked up a small section and held it up to the light. I was staring at frames from

The Battleship Potemkin. The silence of this deserted city was only broken by the noise of birds singing in the bushes sprouting from the long-neglected flowerbeds.

Beyond Chagan the tarmac road occasionally gives way to a muddy dirt track. There is no money for repairs. The remote and arid steppe across which Genghis Khan marched his vast army was once the haunt of nomadic farmers. But in 1947, that all changed. The territory was chosen by the Soviet defence ministry as their nuclear test site. Tens of thousands of workers poured into the area, which was quickly transformed into one of the richest parts of the Soviet empire. By 1949, the huge construction programme was complete. Roads, railways, water supply conduits, power and communication lines, towns and cities were built to a high technical standard. A sophisticated infrastructure was put in place to measure the atomic blasts around Ground Zero.

It was at the end of the line on the Trans-Siberian Railway that Josef Stalin ordered a new town to be built in 1946. On the remote plains of Kazakhstan, the USSR's leading physicists began to assemble. Their top-secret task was to build Stalin an atomic bomb. Three years later, Kurchatov – the town named after the nuclear scientist who led the project – fulfilled its promise to Stalin.

At 7 a.m. on 29 August 1949, near the village of Dolon, the Soviet Union detonated its first atomic bomb. It was to be the first of those 607 nuclear devices secretly exploded at Ground Zero over the next four decades. Further massive above-ground explosions in 1951, and the first plutonium bomb in 1953, followed. On 12 August 1953, the first test was carried out on the new type of atom bomb designed by Andrei Sakharov. This device, based on Sakharov's innovative 'layer cake' design, involving layers of fusion material placed into concentric shells of an enlarged implosion device, was exploded in the Semipalatinsk Polygon, creating an explosion which measured 400 kilotons. Sakharov says in his memoirs that he was nervous and had taken a sleeping pill and gone to bed early the night before the test. At four in the morning he and the rest of his team were awakened by alarm bells and set off on the two-and-a-half-hour journey to the control centre, located around 20 miles from the bomb.

In his memoirs Sakharov describes the moment of detonation:

'We saw a flash and then a swiftly expanding white ball lit up the whole horizon . . . I could see a stupendous cloud trailing streamers of purple dust. The cloud turned gray, quickly separated from the ground and swirled upward, shimmering with gleams of orange. The customary mushroom cloud gradually formed, but the stem connecting it to the ground was much thicker than those shown in the photographs of fission explosions. More and more dust was sucked up at the base of the stem, spreading out swiftly. The shock wave blasted my ears and struck a sharp blow to my entire body; then there was a prolonged, ominous rumble that slowly died away after thirty seconds or so. Within minutes, the cloud which now filled half the sky turned a sinister blue-black colour.'

A colossal thermo-nuclear device was dropped onto the site from an aircraft in 1955, sending a radioactive cloud across most of Kazakhstan and into China. More than 600 nuclear explosions later, and following sustained protests and peace marches by the courageous Kazakh people, largely ignored in the West, the Soviets were finally forced to abandon plans for further tests in 1990.

Kurchatov – 'The City of Angels'

The city of Kurchatov, 150km from Semipalatinsk, was home to over 30,000 residents, including scientists such as Andrei Sakharov and Stalin's notorious KGB chief, Lavrenti Beria. Stalin had ordered Beria to execute Sakharov and Kurchatov if the first atomic test failed. Today only 9,000 people live in Kurchatov. Most of the scientists who remain are engaged in the study of radiation and nuclear safety. There is mass unemployment and a tangible air of despondency. Like elsewhere in the Polygon, the city is crumbling.

It was the elderly Professor Balmukhanov who told me why Kurchatov had come to be known as the 'City of Angels'

during the Soviet era. A little peasant village belonging to a
poor Soviet collective farm (*sovkhoz*) called Kzyl-Zhuldys
(Red Star), it is nestled against a steep bank of the River Irtysh
deep inside the Polygon, near Kurchatov. There were only 30
or 40 houses here in the 1950s, but the whole village was more
than 1km long. Each house was surrounded by a large cattle
yard with rudimentary corrugated tin byres.

One evening, everybody was waiting anxiously for the old
cowhand to come back from the pastures with the village cows.
The women of Kzyl-Zhuldys were expecting to milk their cows
and use the milk to boil a fistful of barley, or, for the luckier
ones, some wheat, providing lunch and dinner combined for
the whole family. But there was no sign of the cows or of their
elderly keeper, who was called Aktykoz. The women were
becoming increasingly grumpy, cursing Aktykoz, while others
tried to calm them down, reasoning that he'd probably taken
the cows to further pastures where there was better grass.
Suddenly, they heard the cows. The herd had found its own
way home, but there was no sign of Aktykoz. A search party
was sent out into the steppe to look for him, but to no avail. He
had disappeared. It was some weeks later before he returned
with his strange tale and the legend of the 'City of Angels' took
root.

Apparently Aktykoz had been tending to his herd of cows
when he was spotted by a mobile military unit from Kurchatov,
whose job it was to protect the top-secret nuclear test site in the
Polygon. They regarded the old cowhand as a suspicious
person. He was arrested and taken to Kurchatov. While checks
were made on his identity, he was held in the military jail.
However, young soldiers, conscripted from nearby villages, felt
sorry for the old man. They gave him extra food and never
locked him up at night like a criminal. One night Aktykoz took
advantage of his situation. Escaping from his cell, he crept
along darkened streets until he reached the city limits. As dawn
broke, he recognised an old cemetery two or three kilometres
away from the nearest military checkpoint and decided to hide
there. When he approached it, however, he found to his
surprise that the cemetery had been desecrated, and all of

the graves had been removed and replaced by a makeshift army barracks.

Soon, he was picked up by a patrol and taken back to Kurchatov jail. However, the conscripts decided not to report him to their commandant. They were fond of the old man and admired his bravery in trying to escape. He asked them what had happened to the old cemetery. Where had all the bodies gone? They told him that his temporary hiding place was all that remained of an old barracks built by political prisoners brought from all over the country to construct the Polygon nuclear site. The prisoners were then taken away and, rumour had it, they were shot and buried inside tunnels blasted into a nearby mountain range.

'But where has the cemetery gone?' asked Aktykoz.

'There no longer is one,' came the reply.

'And where do you bury your dead people?'

'Nobody dies here, and those who reach 50 just fly away like angels.'

'They fly away? But where to?'

This question from Aktykoz was met by silence from the soldiers, some of whom glanced upwards, towards the sky.

It was only years later that the truth about the City of Angels leaked out. Indeed it was true that Kurchatov did not have a cemetery, and nobody was buried there, on the orders of a paranoid Stalin. The ill and the dying were taken to a city in Ukraine, called Yellow Waters, where they were buried in a cemetery protected and patrolled by the KGB.

In 1989 Professor Balmukhanov questioned Colonel Turapin of the KGB about the real reasons why there was no cemetery in Kurchatov. He said it was not much of a secret. The bones and teeth of the dead retain plutonium, and if properly measured, it is possible to get an accurate idea about the size and strength of the nuclear devices which had been detonated in the area. Colonel Turapin said that there were always fears about possible spies trying to dig up the dead, so Stalin ordered that no one should ever again be buried in Kurchatov.

As you leave Kurchatov, a few kilometres on from the last

former Soviet army checkpoint, the tarmac ends and the journey to Ground Zero continues off-road, across the parched and endless steppe. Despite the searing heat, vehicle windows have to be kept tightly shut to avoid inhaling plutonium particles in the swirling clouds of dust. Together with my interpreter Elena and Dr Kamila Magzieva, my Kazakh friend and adviser, I was packed inside an old minibus, perspiring in the sweltering heat. A young scientist, equipped with Geiger counter, a press photographer and a driver completed our small party. Soon, a spiral of dust could be seen approaching fast across the steppe. It was a local villager riding an old motorcycle/sidecar combination, hurrying to escape arrest for pilfering copper wire and metal from Ground Zero.

The Kazakhs do not have the resources to police the test site, and, despite the fact that spending more than ten minutes at the epicentre is lethal, many villagers camp on the site for days, digging up the hundreds of kilometres of copper wire used to detonate the bombs. They know they will die in a few years from radiation poisoning. But they say they will die anyway from starvation. At least, they argue, this way they earn enough to feed their families by selling the copper across the border to the Chinese. The problem is that this deadly radioactive copper is then fashioned into jewellery and sold in China or exported to the West.

At 5km from Ground Zero, the first series of reinforced concrete towers, known as 'geese', still bearing nuclear blast monitoring equipment, can be seen. Nearer Ground Zero, the towers are little more than mangled heaps of steel and concrete. Rocks and stones have been turned to glass. The eerie stillness of the place belies its former hideous purpose.

Here camels, sheep, pigs, cattle and dogs were tethered to stakes to await the scorching nuclear blast. A whole small, uninhabited town was erected nearby with dozens of wooden houses, two brick-built shops, an exact replica of a train station, a factory, and road and railway bridges. Scarecrows dressed as soldiers were dotted around. Military machinery, artillery pieces, tanks, planes, transport vehicles and armoured cars were placed at different distances around the epicentre to study the impact of the bomb. Now the tangled detritus is all that remains. The shrill bleeping of a Geiger counter breaks the silence. A lizard rustles in the undergrowth around the rim of the massive crater. Locusts hop aimlessly from plant to plant. The ground shimmers in the heat.

The young scientist looks at his watch. 'You've been here for more than ten minutes already,' he says. 'We must go.' The photographer is still taking pictures of the crater. We shout to him to hurry. We are all being exposed to dangerously high levels of radiation.

Hours later, back in Kurchatov, we are told to throw away our shoes and clothes and to wash thoroughly in a shower. This was easier said than done. The one available shower managed barely a trickle of water. 'Drink plenty of vodka,' our friendly scientist advised. 'It flushes the radiation out of your system. If you don't, you will feel really ill tomorrow.' We took his advice and drank lots of local Kazakh vodka – and still felt really ill the next day!

The next morning we met Melgiz Metov, who describes himself as a 'nuclear' soldier. He is the chairman of the Council of Veterans of the Semipalatinsk Polygon. He first came to the Polygon 44 years ago as a conscript within the Electromagnetic Emissions Department of the Engineering Corps. He remembers the test carried out on 6 August 1962. 'The rocket carrying a nuclear charge was meant to explode in the atmosphere, but instead it went up for a bit and then down until it hit the ground. So there we had an unplanned ground explosion. But the worst thing was that the wind was blowing towards us, and the cloud of dust and debris descended on us.'

Soon after that Melgiz became very ill, suffering from hypertension and nervous ticks. He had to retire at the age of 35 on invalidity benefit, and then had a serious operation. When he left the Polygon, he wanted to erase any memories of the place, where a lot of his comrades had fallen ill and died. 'I can't believe I survived all this, because most of those who worked here died a long time ago when they were still very young,' he says.

Znamenka

The village of Znamenka lies in the heart of the Polygon. It was one of the worst affected villages. It is a typical ramshackle Kazakh affair, with mud-bricked and grass-roofed huts, a baking 40°C in summer and a shivering −40°C in the snowbound winter of the steppe. This would be an unwelcoming place to live at the best of times. But now is the worst of times. The departure of the Soviets in 1992 led to economic collapse. An attempt by the Kazakh authorities to privatise the old system of collective farming failed. There is high unemployment and no job opportunities. There is also the legacy of the Cold War – Stalin's legacy.

The village elders tell their story to anyone who dares to visit. Unlike their Russian-speaking neighbours from the city, they still speak Kazakh. Many remember the ground shaking beneath their feet and the mushroom clouds rising in the distance. They were encouraged to come out of their homes to watch. The authorities told them they were privileged to witness the might of the Soviet military machine. They were not told that many bombs were detonated only when strong winds could ensure a thick cloud of radioactive dust would blow in their direction.

The first time I visited Znamenka the women of the village had gathered in the school to await the arrival of our party. I was accompanied, as usual, by Elena Kachkova and Dr Kamila Magzieva, and also the *akim* of Semipalatinsk and the Minister of Education from the Kazakh government, along with various other civil servants. The school is the only three-storey building in the village. Built in Soviet times, it caters for more than 500 children. The head teacher told us that their entire budget for the whole of the previous year amounted to only $160. Even so, the villagers had somehow managed to paint the classrooms and hall in anticipation of our visit.

Fifty or sixty local women had gathered in the main hall of the school. They had come to tell their stories to the foreign visitors. They explained that everything is contaminated – plants, animals, insects and humans. Radiation and salt have polluted their only source of water. They are forced to eat the few sickly cows and sheep that remain. Nearly every woman in the room was visibly ill.

A 38-year-old said her breast was removed last year due to cancer, but she was lucky to have found a job, and must work to live, as her husband had died of cancer. She looked pale and sick. An old lady was helped to her feet. She explained that her joints were stiff and crippled – a common ailment in the Polygon. She was sure it was due to the radiation. Her husband had died of cancer two years before. She couldn't walk and couldn't work, and had no one to turn to for help. Despite her appearance, she was only 48 years old. Premature ageing is another common feature. The tears rolled down her cheeks as her friends helped her back to her seat. A big lady in a tattered dress summed up the mood of the meeting. 'All we need is clothes to wear and food to eat to be like anyone else in the world.'

Across the street in the village medical centre, the local doctor and nurses were struggling to cope. Often these dedicated people go without pay for weeks. There is little money for basic medicines, and no money for equipment. They have to deal with all the usual medical problems of a remote rural community numbering 4,000 people, but in addition they have the cancers, birth defects and illnesses caused by the nuclear tests. They work in a ramshackle shed with a tin roof. An old broken fridge acts as a medicine cabinet.

The doctor explained that she had 70 patients whose medical conditions were directly attributable to the nuclear legacy. However, the state authorities demand a rigorous series of tests over many months, and sometimes years, before they will provide a certificate accepting the patient as a radiation victim. Such certificates entitle the victims to a tiny weekly payment and free medicines.

An elderly mother brought her son into the room. Like many of his peers, he had severe learning difficulties. He was 21, but had the mind of a four-year-old. He suffered from epileptic convulsions, and years ago the local doctor had prescribed a specific drug she knew would help. The mother wept as she described her frustration. She waved the old, crumpled prescription in her hand. The authorities had still not classified her son as a victim of the bomb tests and she could no longer afford to pay for his medicine. The doctor said that only seven out of the 70 local radiation victims had been classified.

We were introduced to Berik Sadykov. He was one of those more severely affected by the consequences of nuclear testing in the region. This 32-year-old man was born with serious birth defects. He had been to Italy to undergo two operations to remove a huge growth on his face. But after two years, his tumour had returned, covering his entire face. He had now lost eyesight in both eyes. Said Berik: 'The Polygon affected me in a bad way. I have a serious illness and there is no cure for it. I have one dream now. I just want to regain my eyesight, even in one eye only, but there is no one out there to help me.'

Mirzagali Asrepov had also lived in the Polygon for more than 30 years. He used to live in the village of Kainar near Ground Zero. Now he thanks God that he is still alive and that he could move his family away from the centre of the Polygon and closer to Semipalatinsk, to Znamenka. He said, 'I remember the year 1965. My father left the house one evening to take the cattle back to the shed for the night. While he was out, we heard a thunderous explosion. My father came back with his face blackened from the ash, he looked very frightened. He very soon became ill and died. He had acute radiation poisoning. I also received a high dose of radiation that day. Now I can't walk, my eyesight is gone. I can't name all my other ailments – there are so many. This is all thanks to the Polygon.'

When we left the crumbling clinic, we could see a small group of village elders – all men – had gathered on the opposite side of the dusty track awaiting our arrival. The elders looked extremely stern. One old and wrinkled man with a brightly embroidered cap and two glistening, gold-filled front teeth drew himself up to his full height and launched into a short lecture. 'We get frequent visitors from the West,' he said. 'Politicians who come to gape and stare and promise to help but are never heard from again. We hope that you are not one of these "disaster tourists".'

It was this encounter with these dignified and wise village elders of Znamenka which more than anything else made me determined not to become a 'disaster tourist' but to provide some tangible help to these beleaguered people. It was this first encounter that drew me into more than a decade of exploration of the social and environmental horrors inflicted on the people of Central Asia by the Soviet Union. The Polygon was simply one of the dreadful legacies bequeathed to his people by Joseph Stalin and his successors.

Sarzhal

The village of Sarzhal lies in the middle of the Abai District of Semipalatinsk. It was almost in the centre of the nuclear test site.

Here, wild horses can be seen drinking from polluted lakes. Kazakh herdsmen on horseback tend their flocks of goats and sheep in the searing heat. Sarzhal was only ten miles from Ground Zero when the first nuclear tests were carried out. Later, the Soviet authorities moved the entire village to 25 miles from the epicentre. Illness and disease have cut a swathe through the local population.

Today, the villagers say that they have suffered enough and they do not want to show their misfortunes to the world any more. They do not believe that anyone in the world cares, or is able to help them, so when they see new visitors in the village, they go back to their huts and close their doors.

Many have ill relatives at home. Serikaisha Madenova used to invite visitors to her house, but today says she can't do it anymore. 'I had thirteen siblings in my family. Today, I am the only one left. My brothers and sisters died because of this contaminated Polygon. I have an ill son at home, he is different from other people, he has stunted growth syndrome, but I am not going to show him to you. And other people in Sarzhal will not show you their children. We are sick and tired of being laboratory animals, and there is nothing for us to gain from talking to you. You must go home, and we do not want you to come back here. Nothing can be made better on this contaminated land.'

According to the elders, a lot of suicide victims are buried in the village cemetery at Sarzhal. Why do the villagers resort to taking their own lives? No one knows the answer. Says the village GP Dr Bolat Serikbayev: 'We just don't know what to do about this. People here often kill themselves, even the young ones. We suspect that this is the influence of radiation, one of the Polygon's secret weapons. People have some kind of psychological breakdown. Suicide is our main problem.'

I was invited to address the local inhabitants of Sarzhal in the village library. At this meeting, the village elders vented their fury at the Kazakh government's failure to provide adequate help. One tall gentleman, wearing a traditional Kazakh embroidered cap, roared his disgust, fingers jabbing the air. He shouted, 'The government will not be happy until we are all dead and the problem has disappeared forever.' He pointed through the window in the direction from which the nuclear holocaust came and recalled the horror of the bomb blasts.

Another man of 80 came to the lectern. He was a decorated war veteran who had served his country at the battle of Stalingrad. In a dignified and quiet voice, he explained that only two years ago he was a happily married grandfather with ten children and grandchildren. Now, only 24 months later, owing to the prevalence of the disease in Sarzhal, his wife was dead from cancer, and eight of his children and grandchildren had also died from the disease. Of his two remaining grandchildren, his eldest granddaughter passed her business studies diploma in Semipalatinsk only last year, then committed suicide, overwhelmed by the tragedy engulfing her family. He said that he had been forced to witness the first thermonuclear test.

A middle-aged woman began to sob quietly at the back of the hall. An elderly man wiped tears from his cheeks. 'How can we live on a pension of 8,000 tenge [$55] a month?' he asked, referring to the special pension given to victims of the nuclear tests. On cue, the sky suddenly darkened and the library trembled as thunder roared across the steppe, almost as if the nuclear tests had begun again. A torrential downpour rattled on the corrugated roof, echoing the tears flowing inside.

Kainar

In the village of Kainar, among the foothills of a low mountain range, villagers in national Kazakh costume had gathered outside a yurt to welcome our group. Salty chunks of dried, curdled yoghurt were offered together with large wooden bowls filled with soured mare's milk. A sheep had been killed in our honour, and I was asked to slice meat from the boiled head, which sat forlornly on a wide dish, horns attached.

Traditionally the ears, being the greatest delicacy, must be cut off first and offered to the most honoured guest. Then slivers of meat from around the mouth and nostrils are cut and served in turn to each guest crouched at the low table. Endless toasts are offered, washed down with mare's milk or vodka. The wise choose vodka!

Soon the rest of the boiled sheep arrived, pieces of carved meat lying on alternate layers of thick yellow fat and pasta. Equally fatty horsemeat followed. The Kazakh villagers must survive temperatures of $-40°C$ in winter, and fat plays a large part in their daily diet. A lack of refrigeration to deal with the searing heat of summer means that milk and yoghurt must be soured and salted to survive. However, radiation has penetrated every layer of the food chain. The water supply is polluted, milk and meat are irradiated and vegetables absorb radiation from the soil. Not exactly haute cuisine.

It was pitch-dark and we were exhausted. Our hosts assured us that the local village 'hotel' had been prepared for our visit. We were bemused to hear that a remote village on the steppe could boast a hotel, but our visions of a Las Vegas-style desert inn with luxury bedrooms and high-pressure showers were quickly dashed. Driving up a dark and dusty Kainar side street, we shortly arrived at a walled enclosure behind which, we were told, a warm welcome awaited us. Tripping over abandoned car tyres and other assorted junk in the dark yard, we picked our way to the door of a long, tin-roofed hut. A young Kazakh man opened the door and beckoned us in. We were standing in the hallway of the village clinic, which had been emptied of patients prior to our arrival. This was the only place big enough to accommodate us, and clearly the poor patients had been turfed out to make way for the foreign guests.

Our young guide pointed out the facilities. A metal basin hung precariously on one wall of the hallway. Above it, an old tin petrol can dangled from a piece of rope. Our guide tilted this ingenious device to demonstrate how water trickled from the petrol can into the basin beneath. This was for washing, he explained. Kamila, Kimberley and Elena were shown to a large dormitory, where filthy mattresses adorned a handful of rusty iron beds. I was privileged to have my own single room, the only furniture being the iron bed, with a mattress that looked as if Genghis Khan himself had urinated on it as a child. A pungent odour hung in the air. Our guide told us that there was an outside toilet at the far end of the garden, but we decided not to risk falling into it in the dark. In any case, we had been invited to visit the village sauna in order that we could decontaminate ourselves after our day in the Polygon, so we were hoping that better facilities might await us there. We dropped off our bags and got a lift to the sauna with the village *akim* in his battered Lada.

Several dark, deserted streets later, we pulled up outside another rickety shed. This one had smoke and sparks belching from a tall brick chimney, clearly signifying the existence of a sauna. Of course, there was no electricity in the sauna, or indeed anywhere else in Kainar. Kamila, Kimberley, Elena and I stumbled into the dark shed and lit several candles. The smell of wood smoke filled the air, and searing heat pulsed from the sauna room. Despite the almost pitch darkness, the three women decided to undress in the actual sauna room itself, while I used the tiny foyer.

Suitably wrapped in an old towel provided by the village *akim*, I was about to knock on the sauna door to establish whether it was safe to enter, when the outside door burst open to reveal a young, wild-looking Kazakh man. He was gesticulating and jabbering madly in Kazakh and was clearly drunk. Before I could stop him, he had pushed his way past me and thrown open the door to the sauna room. He was met with squeals of dismay from the women, followed by a torrent of angry Kazakh from Kamila. The drunken man, it transpired, was in charge of our sauna. He had been asked to set the fire and get the sauna working late at night for us to use and, as a reward, had been given a litre of vodka. He had obviously consumed the entire bottle before deciding to pay his respects to the almost naked foreign visitors. As we sat sweating out radiation in the darkness, we could hear our drunken friend singing loudly behind the sauna as he piled more and more logs into the roaring furnace.

Our sauna adventures were not quite over either. We found a huge barrel of water poised precariously on the flat roof of the building, from which protruded a short hose. By pulling out a plug from the end of this hose it was possible to get a fairly rapid trickle of cold water in the form of a sort of shower. One by one we discreetly showered in the darkness under this DIY device. We dried ourselves and got dressed before stumbling outside to search for the village *akim* who had promised to come for us in his Lada. There was no sign of life. Even the drunken sauna attendant had disappeared. In the distance a dog barked. A million stars sparkled in the great steppe sky. But where were we? We knew our so-called hotel was only a few streets away, but the streets are unpaved, unnamed and unlit. We hadn't a clue which way to go. An hour passed. We were beginning to feel the cold of the night creeping into our freshly saunaed bones when at last we could hear the sound of a car approaching. It was the *akim*. He apologised profusely for being late. Never have four people clambered into a Lada with such enthusiasm and soon we were back in our 'hotel', where a fitful sleep awaited us.

Next morning we visited the cemetery just outside Kainar, which is almost bigger than the village itself. Grave after grave bear the pictures of young men and women, victims of cancer or suicide. The inscriptions are poignant. One young woman died at the age of 20. Her name was Orazken Malkarbay. On her tomb is written: 'She did not reach her twenty-first spring and left us suddenly. Crying forever. Her Father.' Our guide explains that 'suddenly' is a Kazakh euphemism for suicide.

Soon we arrived at the village hall in Kainar, which was filled to overflowing. More than 500 people had turned out to greet us and tell us of their suffering. The people of Kainar are highly articulate and never slow to grasp an opportunity to air their views. They were particularly aggrieved by the fact that although they had been promised a new water supply piped from clean, un-irradiated springs in the mountains, it had still not materialised. Instead the government was talking to them about a project to build a museum in the village. This infuriated them.

Some heated exchanges began between the villagers and the local politicians. By now we were three hours behind schedule. Sixteen scientists from the National Nuclear Research Centre in Kurchatov were waiting for us at the Atomic Lake. They had brought protective clothing and gallons of water to wash us down after our visit. However, our guide had a better idea. He had agreed to a suggestion from a villager that we should take a shortcut across the steppe, cutting our journey time to the Atomic Lake in half. But first, even though we were desperately late, we had to submit to the usual village hospitality.

Hot and exhausted, we were taken to the *akim*'s house for the final village banquet. The ubiquitous sheep's head was waiting for me at the head of the table, its teeth bared in a rictus grin. The toasts and pleasantries over, we set off in a convoy of vehicles across the grass-covered plains, dust billowing behind us.

The journey by road should have taken just under two hours. After four hours bumping across the prairie, we realised we were lost. Soon we spotted a small ridge rising from the plain and made our way towards it, hoping to get a better view of our surroundings from the summit. The ridge had a broken fence surrounding it, which should have sounded some alarms for us, but it was only when I got out of our Land Cruiser and walked to the top of the ridge that the full horror of our situation dawned on me. I was staring into an atomic bomb crater!

We had inadvertently stumbled across one of the nuclear bomb test sites – potentially lethal and where all access is strictly prohibited. Dr Marat Sandybaev came running up, waving his Geiger counter. 'It's registering 160 Roentgens – a lethal dose,' he shouted. 'We have to get out of here quickly.'

We set off again at high speed, bouncing across the uneven terrain. After an hour, we stopped for a comfort break when suddenly I noticed smoke billowing from underneath the Land Cruiser. I leapt out of the car shouting 'Fire!' at Kamila, Kimberley and Elena, who were squatting side by side in the grass, relieving themselves. Prairie grass had wound itself tightly around the drive shaft and ignited against the hot exhaust. Our driver dived under the vehicle with a cloth. I threw bottles of water to him. The flames were licking dangerously close to the fuel pipe and already the tall grass beneath the car had caught fire. Kamila, Elena, Kimberley and I beat out the grass fire using rubber car mats pulled from the Land Cruiser. For five minutes, the driver fought the blaze under the vehicle, finally emerging blackened with smoke, his right hand and forearm severely scorched. He had almost certainly saved our lives.

Around 9 p.m. we found a Kazakh herdsman on horseback and asked him for directions. He told us to follow a distant line of broken poles, which once brought power across the steppe to the nuclear test sites. After another hour we found the crumbling township which once housed the Soviet military guards and KGB personnel. Our Geiger counter was still recording abnormally high levels of radioactivity.

It was past midnight before we finally discovered an asphalt road and headed for a small village, where we were able to awaken the owner of the only petrol tank for miles around. We refilled the Land Cruiser and headed back towards some kind of civilisation. Unbelievably, as we neared the city of Semipalatinsk in the wee small hours of the morning, we suddenly came across two cars waiting for us at the roadside, with a small feast of caviar and vodka laid out, picnic-style, to celebrate our survival. Word had somehow been sent ahead, and the Kazakhs, much relieved that we had not disappeared forever in the endless steppe, were determined to drink to our adventure, no matter what time it was!

Next morning, all of us were sick. Dr Marat Sandybaev had

to take a bed in his own hospital. He was on a drip with severe vomiting and diarrhoea. We had all been seriously overexposed to radiation. It was a sharp reminder of the ever-present dangers of the Polygon, unseen yet deadly.

Karaul

Our next village visit in the Polygon was to Karaul in the Abai district of East Kazakhstan, named after the great Kazakh poet and humanitarian Abai Kunanbaev. It was Abai who translated the works of Robert Burns and Robert Luis Stevenson into Kazakh. It seems to be the ultimate irony that Stalin should chose the home of this national icon who wrote about love and humanity as the site of his nuclear tests.

In the medical centre we were ushered into the room of a beautiful 14-year-old girl called Aigerim. She stood as we entered. She was wearing a fashion T-shirt with 'Love 7' emblazoned on the front and a pair of flared jeans. She had incredibly sad eyes. The chief doctor explained that, like all other children in the area, Aigerim had chronic anaemia. However, they had been unable to get her blood back to normal and she now had chronic hepatitis, kidney failure and the onset of scoliosis – the condition where the spine can no longer bear the weight of the head and begins to bend painfully. Aigerim listened to our expressions of sympathy, her sad eyes telling us that she only yearned to be like any other teenage girl, away from this place of pain and suffering.

Aigerim's proud silence preyed on my mind for the rest of the visit to Kazakhstan. Her desperation masked by steely determination was so real I could almost feel it, and it served only to strengthen my resolve to do even more to try to alleviate the ongoing suffering and despair these people are forced to endure.

Urdzhar

Kazakh entrepreneur Nurlan Kapparov and his charitable Zhambyl Foundation have refurbished a former cooperative farm and turned it into a new school for handicapped children in the village of Urdzhar. Mr Kapparov has ensured that the running costs of the new school are co-financed by the Semi-palatinsk Oblast and the Zhambyl Foundation. Together they can provide attractive, above average salaries which can help with the recruitment of excellent medical and teaching staff who are able to provide a high level of care and education for the school's pupils.

Having agreed to help Nurlan Kapparov to fund the school for handicapped children, I had been keen to examine the project at first hand. The problem was that the runway at Urdzhar had collapsed into disrepair and there had been no flights to the village for 11 years. Urdzhar is only 50km from the Chinese border but is seven hours by road from Semipalatinsk. Knowing the state of the roads, I had been reluctant to undertake the journey unless suitable transport could be found. The end result was that the local *akimat* (city hall) in Urdzhar invested in the refurbishment of the runway and airport, and persuaded the domestic air service – Semey-Avia – to inaugurate a new thrice-weekly service between Semey and Urdzhar.

Indeed it turned out that when Nurlan Kapparov invited me to visit the school in Urdzhar for the first time, our flight was to be the inaugural flight, the first to Urdzhar for eleven years. An old Yak-40 had been specially chartered for the occasion by Nurlan Kapparov's sister – Sandugash Kisembaev – head of the Zhambyl Foundation. Nurlan and Sandugash's mother had been born in Urdzhar, so there was a strong family connection to the area.

Our fellow passengers for the chartered flight were the *akim* of Semey, a film crew and journalists who were coming to mark both the reopening of the airport and my visit to Urdzhar. The pilot was none other than the head of Semey-Avia himself. I was ushered down to the front seat in the old Soviet warhorse, next to the *akim* of Semey. The heat inside the plane was intense. Although the cockpit is tiny, three pilots are required to fly these ancient jets, and they duly squeezed past me to shut themselves into their cramped accommodation.

The engines roared into life and we began to taxi down the runway.

Suddenly the door opened and one of the pilots emerged. He leaned over me and tried the handle on the emergency exit. He opened the cockpit door and there was a brief exchange in Russian with the other two pilots. He then lumbered to the back of the plane and did the same again at the rear emergency exit. Once more there was an exchange of views in the cockpit. This time he leaned right across my seat and, although we were still taxing along the runway, wrenched open the emergency exit and quickly slammed it shut again. This time there were grunts of approval from the cockpit. He disappeared back inside and shut the cabin door behind him. The old Yak slowly turned and faced east. We roared down the runway and lurched into the air just as the tarmac ran out and the perimeter fence approached.

Halfway through the flight, with 'No Smoking' signs in Russian above each seat, the distinctive smell of tobacco smoke drifted out from under the cockpit door. The pilots were having a fag! God knows how they managed to breathe in such a tiny cockpit with three of them smoking. An hour later, the senior pilot announced that we were beginning our descent into Urdzhar. I could see a cluster of buildings below, but no sign of a runway. Our plane circled several times. 'I think he's searching for the airport,' I told the *akim* sitting next to me. Several of the passengers were by now on their mobile phones, perhaps asking local friends if they could shed some light on the whereabouts of the runway! At last we found it, but still the pilots hesitated. They circled the airport twice more, cautiously checking the state of the newly tarred runway and the best way to approach. Finally, the wheels came down and in we went, coming to a halt with our nose over the grass at the far end of the new tarmac, despite full reverse thrust after we touched down. Clearly the runway was a bit short for a Yak-40. I nervously pondered whether it was long enough for us to take off again later in the afternoon, or whether we were here for the weekend.

A big welcoming party was waiting for us on the apron. Girls in national dress were there with the usual gifts of sweets and sour *kumis*, or fermented mare's milk. The local *akim* and his team were waiting to greet us, and a large contingent of journalists joined up with the camera crew who had accompanied us from Semey. I did some interviews, then we were whisked straight to the existing school for handicapped children, where staff and kids were lined up waiting to cheer us in.

Inside the main classroom, three young children in wheelchairs burst into song as we entered, accompanied as always by an elaborate sound system and DJ. We handed over the bouquets we had received at the airport to these kids and were then shown to a table and chairs at the end of the classroom. The head teacher said that she wanted to tell the story of the school by showing us slides projected onto a large screen. We would see the children perform dances and songs, and hear the accounts of some parents who were also present. She explained that the present school could cater for around 50 children, although there were only 37 pupils at that time. She told us that out of a population for the whole Urdzhar region of around 95,000, there were 780 children born last year, 180 of whom had been born with inherited diseases caused by the nuclear tests. Of these children, 90 per cent would end up as our patients she said, explaining that the school took kids from age 3 to 18.

The head teacher said that she had been told at first that it was pointless to try to educate these children and that it would be better simply to leave them at home with their parents. But she said that the school had successfully developed four separate types of rehabilitation programme, and now had an excellent system suited to local needs. The system mixed education with medical treatment. The pupils are divided by age and degree of disability. They suffer from a range of disabilities, including cerebral palsy and diabetes. Many are wheelchair bound, but with constant therapy and dedicated attention they can often be taught to walk again.

Several kids who performed dances and songs for us had previously been confined to wheelchairs and had now regained mobility. The head teacher explained that the school provides day facilities for local children, although they also have 11 beds for kids from more distant villages who cannot return home each night. She said the current school was her pride and joy, but there was a need for a much larger facility, capable of taking 150 pupils in the first phase and with specialist equipment, staffed by both doctors and teachers. 'There is no future without a past,' the head teacher said. 'Those who forget their past will have no future.'

After these wonderful performances by the children, I heard accounts of how good the school was from some of the local parents. Then we sat in a class for pre-teen kids, some of whom were paralysed from the waist down, watching while they were taught the alphabet. It was striking how happy they all were. There were no tears. There were no grumbles or complaints. Indeed this has always struck me as a common factor in all of the orphanages, hospitals and institutions for children that I have ever visited in the Polygon. These are children who have been dealt a miserable hand by fate, yet the love and affection lavished upon them by their carers has given them a cheery outlook and a disposition that belies their awful circumstances as victims of the Cold War.

A tour of the school craft workshops revealed more spectacularly talented work done by the children. They auction these items every year at a local village fete to raise money for the school. The kids gave me some gifts that they had made, then I posed for a group photo outside the front door of the school with all of the children and teachers, before heading off to see the site for the new school that has been gifted to the Zhambyl Foundation by a local businessman. It is where the former headquarters of a co-operative farm was based during the Soviet era.

A large 1940s building set in around 20 acres of rich farmland and surrounded by gardens and trees, it is situated just on the outer edge of Urdzhar. Already building work had begun. The existing building had been cleared of rubble, and brickies were busily building a new workshop and kitchens adjacent to the main house. We were given a guided tour, and the head teacher explained that the first phase to accommodate 150 handicapped children would cost a total of $650,000 to complete, including all building work, fabrics, fittings and special equipment. She said they hoped that the initial reconstruction would be completed within eight months and the fitting out of the building could begin later next summer.

Although not within the nuclear testing zone itself, Urdzhar suffered the consequences of atmospheric nuclear tests which were deliberately held on days when the wind was blowing towards China. The radioactive clouds from bomb blasts up to 500km away drifted across their village and then, encountering the mountains that divide China from Kazakhstan, settled as rain into the local lakes and rivers. The problem has been further compounded by Chinese nuclear tests, when they deliberately steer their radioactive fallout towards Kazakhstan to minimise health problems for their own local population. Once again, Urdzhar was the recipient of this unwanted and lethal pollution, and once again it was the children who suffered and continue to suffer. I have therefore a great sympathy and support for this project.

Urdzhar was chosen as a health spa by the Soviets for the rest, recreation and treatment of military personnel because of the apparent healing qualities of its local water. It is therefore an ideal location for a school for handicapped children. The village is also lush, fertile and green, and the houses and gardens are tidy and well kept. It is altogether a pleasant environment for such a venture. Its key problem, of course, is its proximity to the Polygon.

Before we could leave for the airport, the *akim* of Urdzhar announced that a lavish lunch awaited us in the main community hall. My heart sank at the thought of yet another grinning skull, the ears and horns still attached and the eyes a pulpy white from boiling. However, as we were led into the hall, where a sizeable party of local dignitaries awaited our arrival, I could smell the pungent aroma of garlic, an unusual ingredient in Kazakhstan. The *akim* explained to me that he had instructed the chef to prepare a 'European dish' in my honour, and they had made a big pot of garlic prawns. I had no reason to doubt this story, as by now the reek of garlic was making my eyes water.

The large table was bedecked with the usual rich array of sweets, nuts, horsemeat, bread and cakes. Sure enough, the ubiquitous boiled sheep's head was ceremoniously carried in, and once more I had to gouge out the eyes and offer them to the *akim*, proposing a short toast along the lines of 'This will help you to keep a watchful eye on your local people.' He gobbled them eagerly. Next the ears had to be severed and proffered to the police chief, with my toast of 'These will help you to keep your ear to the ground to hear what is happening in Urdzhar.' The nose was cut off and given to someone with a toast 'To help you smell the flowers and scents of this beautiful village,' and so forth. By now I was becoming a bit of a professional at this task.

Kamila was standing next to me, providing expert interpretation into Kazakh of all my toasts. Just as I had completed the carving of the sheep's head and launched into a short speech of thanks to the people of Urdzhar, the chef proudly entered the hall carrying the enormous plate of steaming, garlic prawns. The smell was overwhelming. As he set the dish down on our table, Kamila, in mid-sentence, suddenly gagged, and thrusting her hand over her mouth, raced towards the toilets. I stood in awkward silence while loud retching sounds could be heard coming from the Ladies at the far end of the hall.

The chef was transfixed and appalled. The poor man had no way of knowing that Kamila was allergic to garlic and that even the faintest smell of the pungent clove was enough to cause her to vomit. He stood staring down at his garlic masterpiece, wondering what had gone wrong. Just then, Kamila came lurching back down the hall, looking faintly green and pointing at the offending dish as if it were poison. 'Take it away,' she bellowed, breaking the silence . . . 'Take it away!'

The chef scuttled back to the kitchen, carrying his untouched plate of garlic prawns, and Kamila took her place once more at my side. 'Where were we?' she muttered, as I resumed my vote of thanks to a stunned audience.

Back at the newly reopened airport, as I suspected, the crew had decided the runway was barely long enough for a Yak-40. We taxied right to the very end of the freshly laid tarmac, and had our nose over the grass before executing a skilful U-turn and revving up for take-off. This time the pilot had all three jet engines roaring at full throttle as if we were about to lift off from the deck of an aircraft carrier. The old plane trembled and shook, and then, abruptly, the brakes were off and we were hurtling down the runway at breakneck speed, praying that we wouldn't run out of road. Just in the nick of time we soared into the air and got the wheels up before we clipped any treetops. We banked sharply and turned our nose to Semey.

On landing in Semey, the cockpit door opened and one of the pilots raced to the rear exit to get down the stairs before any of the passengers. We soon discovered why. As we alighted he had to steer each of us to one side to avoid fuel dripping onto our heads from the hot rear engine. There was a strong smell of aviation spirit as a puddle of fuel formed on the tarmac and vapour rose from the red-hot engine. No doubt this was a regular problem with the antiquated Yaks, but to passengers more accustomed to Air France or BA, a little unsettling nevertheless. However, this was not to be my last visit to Urdzhar.

Last year I returned to Urdzhar to join Nurlan Kapparov and the *akim* of Semipalatinsk Oblast for the opening ceremony of the brand new school for handicapped children. Thanks to the generosity of Mr Kapparov, whose mother was born in Urdzhar, these children will have a future after all. In fact, there was a rather poignant and wonderful link forged between this remote school near the border with China and a primary school in one of the poorer areas of Glasgow in the west of Scotland.

In early 2010, I paid a routine visit to St Paul's Primary School in the Whiteinch district of Glasgow to talk to the children about my work as an MEP. I told the kids about the Polygon and the children affected by the Soviet nuclear legacy. I also explained that I was going to Urdzhar in June to open the new school for handicapped children. In March the St Paul's kids held a 5km playground marathon and a 'bring and buy' sale to raise £233 for the kids in Urdzhar. They asked me to deliver the money and read out a letter from the St Paul's children at the opening ceremony in Urdzhar, which I happily agreed to do.

Some months later, in my Brussels office, I received a large package containing a heap of photos, a decorated plate, a poem and a letter from the Urdzhar kids, directed to their friends at St Paul's. I contacted the head teacher at St Paul's, who invited me to attend a full morning assembly at which all the 150 children and some parents would be present, and where I could hand over the gifts and show them photos of the opening ceremony in Urdzhar in a PowerPoint presentation.

This event went down so well that I decided to launch a letter-writing competition in which the 10- and 11-year-old children in each school were invited to submit short essays explaining why they would like to meet President Jerzy Buzek of the European Parliament and visit Brussels. A boy and girl winner from each school were selected as winners and, together with parents and teachers, came to Brussels in mid-July 2011. For all of the children and adults, it was a once-in-a-lifetime experience, but for the kids and their parents from Urdzhar, in particular, it was almost as if a miracle had happened. Both of the Urdzhar children were suffering from leukaemia, and both had undergone lengthy medical treatment. Neither had ever travelled further than a few kilometres from Urdzhar.

Their journey to Brussels was a major adventure, involving three days of constant travel. First they had to undertake a 17-hour bus journey to Ust Kamenagorsk. Then they flew from Ust Kamenagorsk to Astana, and from Astana to Frankfurt. Finally they caught a flight from Frankfurt to Brussels. Staying for three nights in a five-star hotel and being shown around Brussels and the European Parliament was for them an eye-opener. For children who had never before seen an escalator or a lift, even these mundane bits of equipment were like Disney-land rides. They were amazed.

The meeting with President Jerzy Buzek was the highlight of the trip. He hugged each of the prize-winning children in turn, and was clearly moved when the two Kazakh kids recited little poems in perfect English in his honour. This was a deeply emotional experience for everyone involved. There wasn't a dry eye in the house!

The competition, inspired by the children themselves and funded by the generosity of Nurlan Kapparov, the Kazakh Embassy in Brussels and the European Conservatives and Reformists Group in the European Parliament, united these two schools through their natural humanity, crossing the boundaries of distance, religion, culture and language to bring these children together in harmony and friendship.

CHAPTER FOUR

Return Visit

On one of my more recent visits to Semey and the villages of the
Polygon, the flight from Almaty to Semey was in a famous old
Russian turboprop – an Antonov 24. These twin-engined
beasts were left behind in Kazakhstan when the Soviet empire
collapsed and were deemed worthless. They are still flying
today. This time I was accompanied by Dr Sin Chai, a close
friend and medical doctor, and my Kazakh 'sister' Kamila
Magzieva. The flight to Semey was surprisingly smooth,
although as we came in to land through a bit of local turbulence,
with the wheels already down and the pilot's voice echoing
Russian commands clearly audible from the cockpit, two young
boys, no more than ten or eleven years old, suddenly lurched
past me, heading for the toilet at the rear of the old plane.

The portly stewardess, who had strapped herself into the
rear seat, almost had apoplexy, shouting at the boys in Russian
to get back to their seats and scrabbling at her seatbelt to try to
break free. From the green appearance of one of the boys,
clearly projectile vomiting was about to take place. I covered
myself protectively with a Kazakh newspaper. By this time
one boy had locked himself in the toilet and the other had been
sent scrambling back to his seat, clutching a sick bag. As the
stewardess banged on the door, shouting at the boy to unlock it
and come out, our wheels hit the runway and the sudden lurch
sent her sprawling, while screams and cries emerged from the
toilet. As we taxied to our stand, the stewardess gathered
herself together and helped the unnerved schoolboy from the
now rather smelly loo. Obviously his first-ever flying experience was going to stick in his memory for years to come.

The usual welcoming party was waiting for us on the runway. As we walked towards them, Sin Chai nudged my shoulder and pointed back towards our trusty aircraft. Two men had climbed a ladder onto the wing of the Antonov, and while one held an enormous funnel, the other poured fuel into it from a large plastic container. Refuelling, Polygon style!

The deputy *akim* of Semey was waiting for us, together with our old friends Zhana from the *akimat* (city hall) office, Mukhtar, head of the children's hospital, and Marat Sandybaev, head of the oncology hospital. They were accompanied by an entourage of local worthies, including two lovely girls in national costume who thrust flowers at Sin Chai and me, and then proffered nuts and sweets and cupfuls of nasty *kumis*. This sour, smelly and slightly alcoholic local delicacy is gulped down with gusto by the Kazakhs, but is certainly an acquired taste. Having pretended to drink a long draught, I smacked my lips in appreciation and placed the still full cup back on the girl's tray. I turned to face the local camera crews and assembled journalists. What is your mission this time? Where are you going? Who will you be giving money to? were the anticipated questions that duly emerged.

Our stuff was loaded into a black Toyota Land Cruiser with a cracked windscreen and we were introduced to Alexander, our driver, whom we later nicknamed 'Schumacher' for his ability to drive in difficult conditions at extraordinary high speeds. We immediately noted that in true Polygon tradition, despite having the worst roads and possibly the worst drivers in the world, the seatbelts in the vehicle had all been painstakingly removed, both front and rear. A police escort sped our convoy into Semey and to the Hotel Nomad, where we dumped our stuff and were then whisked off once more for lunch in the *akimat*.

Inside the old, sinister Soviet council building, Zhana led us to an ancient lift that took us up to the fifth floor, after which we had to walk along a corridor to another lift, which took us back down to the first floor – a cunning KGB design to fool would-be assassins, but probably deadly in the event of a fire. Here, in a room devoid of air conditioning and sweltering in the 40-degree outside temperature was a table groaning with Kazakh fare. Plates of fatty horsemeat were sweating gently in the heat. Flies were buzzing with excitement around some rather limp sardines which had been laid across bits of bread adorned with soft cheese. The table was, as always, covered from end to end with food of all descriptions, virtually none of it tempting to sensitive Western stomachs.

Vodka shots were poured, and the deputy *akim* duly launched into the first toast of the day, welcoming us to Semey and wishing us well on our expedition to the Polygon. I replied, while waving away plates of fat lamb and bowls of fatty soup which appeared from nowhere at regular intervals. I nibbled gingerly at almonds and walnuts, and then took an adventurous bite of a surprisingly delicious Semey tomato.

Lunch over, we took a quick tour of Dr Marat Sandybaev's oncology hospital, visiting the new breast cancer wing and seeing the state-of-the-art equipment they'd acquired. The contrast with the way this place was nine years ago when I'd first visited was amazing. We were shown the model of the proposed new hospital which is yet to materialise, but looks likely to happen, and were then taken into some of the four-bed wards to meet the female patients. I dismantled some of the bouquets of flowers we'd received at the airport and handed single stems to each of the patients.

We moved on to Mukhtar's children's hospital, where again there were signs of improvement, although the cratered and potholed access road into the car park resembled a bomb site; it must be a horrendous experience for really sick children arriving for treatment. However, the corridors and wards had at least been painted, and the floors re-covered. Shortcomings in the building were massively overcome by the dedication and professionalism of the fantastic medical staff. Their real love and care for their patients was quite astonishing. Again we were taken through some of the wards, where mothers sat holding the hands of their sons and daughters, mostly suffering from diseases like leukaemia and anaemia and other common ailments brought about by exposure to radiation.

Our hospital visits over, we set off in the Land Cruiser on the five-hour cross-country journey to the remote village of Kainar. It was already 4.15 in the afternoon. The heat was searing. Even with the air conditioning on at full blast, the windows were still hot to touch. On the perimeter of Semey, next to the huge Muslim cemetery, our police escort peeled off with a wave of the hand. We were on our own. Somewhere behind, Mukhtar and Marat were following in another car.

Soon the properly surfaced road deteriorated into a pockmarked assault course. Alexander swerved at high speed between deep holes and cracks, sometimes jerking the wheel so that we careered off the road altogether and bounced alarmingly along the edge of a ten-foot drop, before swerving back onto the broken tarmac once again. This was his way of avoiding the worst of the hazards. Now and again he was too late to execute these skilful manoeuvres and we would crunch into a deep rut or, worse still, leap into the air off a hidden undulation. After half an hour of this I felt as if I had gone 15 rounds with Mike Tyson and we still had four and a half hours to go.

The wide steppe stretched into the distance. On the horizon, black clouds were occasionally lit by flashes of lightning, as an electrical storm moved slowly towards us. Wild horses grazed unconcerned as we roared past. As the sun began to sink in the west, the spectacular light show made us forget our aches and pains. Great streaks of sunlight speared through the storm clouds. Reds and pinks lit up the sky. A distant line of hills had turned a gentle orangey-pink hue. We were transfixed. The breathtaking beauty of the steppe which had inspired generations of nomad poets and philosophers unfolded before us. It was all too recently that millennia of nomadic life had been rudely interrupted by the brutal oppression of the Communists and the horrors of their 607 nuclear tests.

At last we could see a small group of cars and people on the roadside in the fading twilight. It was the welcoming party from Kainar. The local *akim*, together with the jolly and rotund head doctor, were waiting for us. An elderly lady in national costume offered cups of *kumis* and threw bits of dried mare's milk yoghurt at us in a gesture of welcome, while the village musician sang us a plaintive song accompanied on his two-stringed dombra, the national musical instrument. Mukhtar and Marat rolled up at this point, and vodka was handed around. We were quite glad to have a couple of stiff shots to relieve the aches and pains of our journey.

The village elders explained that they had prepared a feast for us in a yurt which had been specially erected at the foot of a local ridge of hills. We debated whether we should wait for Kimberley Joseph and her film crew, who had been making a documentary in the Polygon for the past month and were due to join up with us in Kainar. But we decided to make our way to the yurt and await her arrival there.

As we drove through the dark and dusty unpaved streets of Kainar, the local head doctor, Akimbaev, explained that the village had had no electricity for four days following a storm. We passed the large cemetery, almost as big as the village itself. Alexander engaged four-wheel drive to negotiate the mucky track, still wet from the recent rain. We bounced and swerved across gulleys and over stream beds, following the lights of the *akim*'s Lada which seemed to power ahead with no need for traction! After ten minutes we arrived at a large yurt. Crowds of villagers were gathered on the grass. It was now almost completely dark, and millions of stars were beginning to appear in the inky black steppe sky.

When our eyes had adjusted to the light, or lack of it, we could see a group of men struggling to start a generator. It gave out an occasional fart or stutter, but seemed reluctant to kick into life. Behind the yurt, around a dozen women had fires crackling away and great samovars of tea boiling noisily. A basin was filled with the legs, ribs, saddle and head of a forlorn sheep, waiting to be boiled. Without a working generator, there was no light inside the yurt, so we all stood around watching the expanding group of men as they fought to get the errant machine started. Loud exclamations were being shouted back and forth in Kazakh as the expert engineers swapped advice. Suddenly there was a bang and a flash of flame. The generator had exploded. Total panic ensued. The older men raced for the safety of higher ground, while the younger men ran to the stream with buckets and basins and proceeded to douse the roaring inferno with water.

While the men shouted instructions from their safe perch some distance above, the teenagers threw bucket after bucket of water onto the blazing generator to no avail. Kamila, who speaks Kazakh as well as Russian, told us that the men were now telling the boys to try covering the blazing machine with blankets. Two carpets duly appeared and were quickly consumed in the flames, which were now getting fiercer and threatening to explode the fuel tank at any moment. One of the boys slung a rope around the fireball and started to haul it towards the nearby stream. Needless to say, the rope rapidly burned through and he landed on his back in the mud. Now, instead of a single fire, there were three separate fires spread across the camp, where bits of blazing generator had broken loose. However, the main part of the burning generator was within easy reach of the stream, and soon copious volumes of water had done the trick and the drama was finally over. Kainar had lost a generator and we still had no light for the yurt.

Around 11 o'clock, beams from the headlights of another vehicle heralded the arrival of a truck from the village with a new generator in tow. Soon this was up and running, and the lights miraculously came on. At long last we were ushered into the yurt, where the low table set at its centre was groaning with food. Stools had been set in a circle around this table and as soon as we took our places, the vodka toasts and dombra music began. Two women then carried in a large tray with the inevitable boiled sheep's head atop a mountain of white mutton fat and scrawny bits of meat. I whispered to Kamila to suggest it be given to Sin to carve. I couldn't stomach having it set down in front of me. My close friend Dr Sin Chai, using his considerable skills as a surgeon, duly severed each ear, the juicy nostrils, the two eyes and the black skin covering the nose and forehead, until there was nothing left but the grinning skull. He speared lumps of this delicacy on his fork and passed them around to all and sundry in true Kazakh fashion. The locals were mightily impressed. The head doctor said that Sin was clearly closer to being Kazakh than me, genetically.

From my stool facing the yurt door, I could see the lights of two cars approaching up the valley. It was Kimberley and her team. They were exhausted, but pleased to see us. I was introduced to Natasha, the co-producer of the film, and to Gavin and Vladimir, the cameramen. Gavin seemed to be a bit the worse for wear, and Elena, my long-time friend and interpreter, who had been part of Kimberley's team for the past four days, explained that he had heard that he had been nominated for an Emmy award for a recent film he'd shot in the Arctic Circle and had been celebrating with a bottle of vodka ever since!

The speeches began again, and finally, when my turn came to propose a toast, in the hope that mine would be the last and we might finally get some sleep, I started with the words 'Allow me to make the final speech of the evening.' To this Gavin was heard to loudly exclaim 'Thank Christ for that' before he lurched towards the yurt door and almost fell over a steaming samovar. But my attempts to bring the feast to an abrupt end were short-lived. More toasts were called for, and more music, until we finally staggered out of the yurt at around 1.30 in the morning. The chief village doctor informed us that we were all to sleep in a wing of his clinic. This news was met with cries of dismay by Kimberley, Kamila, Elena and me. We had slept there last time we were in the Polygon and cynically nicknamed it the Kainar Hilton.

As we headed to our cars, the men and women who had been preparing our feast during the evening could be seen squeezing into the yurt. Their feast was about to commence, at 1.30 in the morning! Soon even the leftovers would be history. We bumped our way back down through the foothills to the darkened village of Kainar. Appropriately, the hospital is located close to the big cemetery. Armed with a torch, the head doctor ushered us to our quarters, a wing of the hospital specially cleared of patients to make room for us. This was the same Kainar Hilton as last time, right enough, although the doctor proudly announced that in our honour they had dug a new outside toilet for us that very day. I borrowed his torch and made my way gingerly up a muddy path until I found it. Really there was no need for a torch; the stink guided me straight to it. I gagged and tried to hold my breath while I relieved myself.

Back in the Kainar Hilton, the other guests were coming to terms with the facilities. A number of multi-bed wards meant everyone had to share, but at least this time the beds had been made up with damp sheets and pillows. The sole sanitary facility for the entire hospital wing consisted of a tin can suspended from the wall above an old basin. When you pushed a knob on the bottom of the tin, water dribbled out into the basin, and from there into a bucket strategically positioned on the floor. Every now and again a young Kazakh woman rushed in with a bucket to refill this antiquated contraption. There was no soap and no towels in evidence.

I shared a room with Sin Chai, which turned out to be a grave mistake. No sooner had I nodded off around two o'clock in the morning, than I was awakened with a start. I thought maybe the head doctor had turned on the electricity generator directly outside our window. In fact it quickly became clear that it was just Sin Chai emitting the loudest snores I have ever heard in my life. It was like a Harley-Davidson bikers' convention!

At 4 a.m. the sun began to rise in a clear blue sky, and at 5.30 I had had enough of the noise of Sin's snoring and decided to wander down through the village. Venturing outside, I met Kimberley and Natasha, who had also had trouble sleeping, and soon we were joined by Sin Chai as well. We set off down the village main street, past rows and rows of what appeared to be shops. The only difference between a shop and a house was the hand-painted sign, usually with a picture and Cyrillic writing announcing the nature of the goods sold within. White stucco walls and grass roofs were offset by brightly painted blue or green wooden doors and window frames. Piles of dried horse dung, to be used as winter fuel, were stacked against the side walls of the buildings. All the shutters were closed and not another person could be seen, despite the fact that the sun was already high in the sky and temperatures were soaring. A few calves and horses wandered in and out between the houses. A cat stopped halfway across the dusty main street to stare at us, unaccustomed to seeing anyone, let alone strangers, up and about at this unholy hour.

After traversing the village, we made our way back to the Kainar Hilton. By now we were desperate for a coffee or tea, or even a bottle of water. Luxuries like a wash, or better still a shower, were distant memories. By now, the rest of the team had emerged from their manky bunks, and we began to pack up our kit and carry the bags around to the hospital car park. Here we were met by a group of patients who were occupying the main wing of the building. An elderly woman rushed forward to grab me in a bear-like hug and smacked a couple of slobbery kisses on each cheek. Kamila interpreted from her rapid Kazakh that she had met me the last time I was in Kainar and was delighted to see me again. 'You come here more often than any of our politicians and you are the only person that represents our interests,' she sobbed. I asked her why she was in hospital and she explained that she had come down with a bout of flu, causing me to take a sharp step backwards.

Other patients came over to chat, and shortly a beautiful young female Kazakh doctor and an equally attractive nurse, both in crisp, clean uniforms, came out to see what was going on. These are trained professionals who could earn ten times their salary if they moved to the West. Yet they remain dedicated to their beleaguered people, working with poor equipment in primitive conditions and for little money. They are the real heroes of the Polygon.

Around nine o'clock, the head doctor arrived and we were once again ushered on board our vehicles for the drive back up the valley to the yurt. Already the catering team was in action, preparing breakfast. Logs were being chopped and fed into the hissing samovar. Cartons of yoghurt and home-made butter were cooling in the mountain stream. A large cauldron was filled with a bubbling mutton stew, clearly the remnants of the sheep we had consumed the night before. My body was screaming out for a coffee. Mutton stew was the last thing I craved!

But before we could take our seats in the yurt, there was an important duty to perform. This was one of the key reasons I had come to the Polygon. I summoned the head doctor from the village and presented him with a large presentation cheque for $5,000. This was money accumulated by the Scottish-based charity Mercy Corps from the sale of my book, *Crying Forever*. So far I have been able to hand out around $120,000 from the book's sales and from sales of some of Kimberley's moving photographs of the Polygon. The money has gone to the oncology and children's hospitals in Semey, and last year to buying an ambulance for the village hospital in Sarzhal. This year it was the turn of the Kainar hospital, and I secretly prayed that the head doctor might use the funds to improve the sanitary facilities.

This ceremony over, the village musician began to sing and play his dombra and the head doctor filled everyone's glass with vodka for the first toast of the day. Thankfully some steaming, milky coffee arrived at this point and I gulped it down, all the while pretending to drink the vodka, while secretly tipping it into my water glass. Later I realised this had been a mistake when I saw Elena draining the glass in one gulp, thinking it was water! She blinked and gasped when she realised she had just consumed at least four shots of vodka! But as someone weaned on vodka in her home city of Novosibirsk in Siberia, I am sure she was none the worse for it.

Breakfast finally over, we rumbled back down to Kainar, where the *akim* had gathered a crowd of around 50 villagers around a platform outside the village hall. He made a long speech about how Kimberley and I were returning to his village for the third time and how each time we visited the Polygon we brought aid and raised the profile of their suffering in the wider world. He wished Kimberley every success with her film and invited each of us to say a few words to the villagers. The elderly lady with the flu whom we had spoken to at the hospital also turned up on the platform, resplendent in a blue uniform coat covered in medals, including the Order of Lenin, the highest award of the former Soviet Union. She explained that she had been a driver during the war years and had received the award for dedicated service and bravery.

By now it was noon and I was keen to get going. We still had to visit the villages of Sarzhal and Znamenka on our way back to Semey, and we had the arduous five-hour drive across the steppe to contend with once again. But the head doctor and the *akim* would not hear of us leaving before we returned, for the third time, to the yurt, so that we could have lunch. So off we set again on the familiar journey up the valley, past the cemetery and into the foothills. Unbelievably, they had killed another sheep, and this time I had to do the carving. 'The Kainar sheep will be bloody relieved to see us go,' I thought, as I hacked off an ear and proffered it to Natasha.

At last, with the plaintive songs of the village musician still ringing in our ears and the fatty mutton lying heavily in our stomachs, we said our goodbyes and headed for Sarzhal and the open road. As we drew out of Kainar and headed for the dirt track that leads back to Semey over several hundred miles away, the toot-tooting of the head doctor's car could be heard behind us. He had chased after us to say one last farewell. He hugged me like a long lost brother, and kissed me wetly on each cheek, telling me that I was truly a member of his family and wishing me long life, health and happiness. It was 2.30 in the afternoon by the time we finally made our escape.

The large, ramshackle village of Sarzhal is home to 3,000 souls. It is situated about halfway between Kainar and Semey. We drove to the local village hospital, hoping to see the doctor and her team and to get a look at the village ambulance that I had funded the year before. A villager explained that an emergency had occurred and the doctor had driven a sick lady to Ust Kamenogorsk for treatment in the ambulance. This was gratifying news, as that is what the vehicle was intended for. We were running behind schedule in any case, so we left our best wishes for the doctor and set off for Znamenka.

In Znamenka, we were met by the village *akim*, a fit and cheery man who looks younger than his 60 years. He knew us from previous meetings and explained that he used to be the *akim* of Sarzhal but last year had moved to Znamenka. He ushered us into the new village clinic which, he explained, was his former home – he had gifted it to the village as a new and much better medical facility. He was now living in a smaller house. This staggering act of charity really put my humble attempts to help into context. Here was a local politician really prepared to put his money where his mouth is.

The *akim* explained that things were getting worse in Znamenka. 'This village used to have 10,000 residents during the Soviet times. Now there are only 2,500,' he said. 'Young people leave for the cities whenever they can. There is no work here. No one will invest in the area. The pollution is too dangerous.' In addition, he explained that there is no drinking water in the village. Clean water has to be transported by truck every single day from Semey, 60km across the steppe. The local water supply is salty and completely unsuitable for drinking or even cooking. They use it only for washing. He said that the villagers also have no access to Kazakh TV. They receive all their TV by satellite and can get no Kazakh channels, so while they have been able to watch me doing my stuff in the European Parliament on Euronews, the international multilingual TV channel largely funded by the European Commission, they don't know what is happening in their own country.

Of course, no visit to Znamenka would be complete without lavish hospitality, and although we were keen to make headway back to Semey, the *akim* insisted that we stay for a cup of tea. This proved once again to involve a table groaning with food and, inevitably, a mutton stew. It seems that mutton must be consumed even for afternoon tea in the steppe! Toasts in vodka were followed by a special performance of a song specially written for our visit by a local poet and teacher, who had put on national costume for the occasion. She serenaded us on her dombra and then was quite overwhelmed when Sin Chai promised to fund the purchase of a number of instruments provided she agreed to teach some of the village children to play. Sin said he was fascinated by this unique cultural tradition and wished to see it preserved and enhanced.

Next the *akim* announced that he had been honoured to place the cape of office on the shoulders of President Nazarbayev at his recent presidential inauguration and he was now equally honoured to provide both Sin and me with similar

Kazakh capes. He unfurled dazzling blue and maroon embroidered gowns and large Kazakh velvet hats, and helped both of us to put them on. He then insisted on handing over his own personal dombra to Sin as a gift. Suitably clad, we now resembled Kazakh knights of old.

At this point Alexander hurried into the room and announced gravely that he had found a nail in the front tyre of the Land Cruiser and that he had only just enough fuel to get us back to Semey, over 60km away, provided we didn't use the air conditioner. He said the tyre would only deflate if he pulled the nail out, so he suggested we should get going as quickly as possible to try to get to Semey before dark. The *akim* summoned a local policeman and instructed him to follow us all the way to Semey in his police car so that we could be rescued if we broke down completely. We set off across the plain, with the sun setting in a wide and fiery glow behind the western hills at our backs. It was 9.30 when we finally reached Semey and the Nomad Hotel. Never was I so glad to see a hotel in my life. Sin and I purchased two large beers and made for our respective rooms. I filled a deep bath and luxuriated in the soapy foam for an hour, while sipping my Budweiser. Luxury!

Ust Kamenogorsk and Astana

The road from Semey to Ust Kamenogorsk in East Kazakhstan is over 200 kilometres long. It has recently been upgraded and is hailed as an example of how the Kazakh government is pouring some of its oil wealth into redeveloping the country's infrastructure. In fact the narrow, two-lane road is bumpy and uneven, as we discovered in late October 2009, when we thundered across the empty Kazakh steppe in a Nissan 4 × 4, sometimes hitting speeds of over 100mph, on roads black with ice.

We were speeding towards the airport to catch our flight back to Almaty. The loneliness of the vast, frozen steppe was broken occasionally by twisted, rocky outcrops. This is the area of Kazakhstan where much of the world's supply of uranium is mined. Sheep were dotted across the landscape.

Lone shepherds on horseback could be seen riding across the frosty plains. A blood-red sun slowly rose beyond the mountains that divide Kazakhstan from Siberia.

It was my eighth mission to this desolate part of Central Asia. I was again accompanied by Kimberley Joseph, the Hollywood actress and photographer who I first brought to Kazakhstan in 2003 and who was now returning for her third visit, together with my usual team of advisers and interpreters – Dr Kamila Magzieva and Elena Kachkova.

To raise international awareness, Kimberley and I have toured an exhibition of her photos of the Polygon around the world, from Edinburgh to Dublin, Norway, London, Brussels and New York. In 2008, we took the exhibition to the old capital of Kazakhstan, Almaty, then on to the new capital Astana, and to Semey, the main city in the Polygon. We even took the display out to the villages themselves to show the people who featured in many of the photos what we were doing on their behalf.

In Astana, a large gathering of MPs had come together in the parliament to listen to a PowerPoint presentation from Kimberley and me. Donations immediately started to flow into a special fund set up in Kazakhstan to raise money for the victims of the nuclear tests. Kimberley told the audience of three or four hundred people that she was astonished to see the marble palaces and government buildings springing up from the desert floor in Astana as the new capital is constructed with no expense spared. She said if even a fraction of this money had been spent on the villages in the Polygon, there would be no problem. The audience were primarily composed of civil servants and government employees. There was stunned silence. You could have heard a pin drop. A row broke out immediately following Kimberley's speech, with a senior official shouting at Kamila, and Kamila, who never takes any prisoners, shouting back!

What we didn't realise was that our speeches had been filmed on national network TV and were broadcast across Kazakhstan. As a result, when we arrived in the Polygon two days later, Kimberley had become a national hero – like Joan of

Arc. Everywhere we went she was applauded by the villagers. The *akim* of Kainar even presented her with a horse, the ultimate Kazakh accolade.

At a meeting with the Kazakh minister of education in Astana, I said that it was ridiculous that his government was not doing more to help. Kazakhstan has vast oil reserves, perhaps greater even than Saudi Arabia, and with oil then fetching over $90 per barrel the Kazakh government had ample resources to provide more aid to the beleaguered citizens of the Polygon.

The minister told me that the government had, in fact, allocated over $200 million in its budget for the Polygon, including a one-off payment of $200 to each person who could provide medical evidence to show they were affected by radiation fallout. But I replied that the villagers had often told me that it was almost impossible to get such accreditation. They needed to provide medical certificates validated by three doctors, after filling in endless forms. Finding three doctors in the Polygon is a test in itself. I told the minister that every time I had visited the villages in the nuclear test zone the people told me that all they wanted was clean water and safe food.

When the area was being used for nuclear tests, the Soviets piped clean water to the villages from the distant mountain ranges. The pipes have long since corroded and collapsed or been uprooted and stolen for scrap, forcing the local people to drink irradiated water from underground springs and streams. Similarly, more work needs to be done to clean up areas of land that could then be used for farming. Instead of this, millions are spent every year on endless consultants' reports. You could almost paper the entire Polygon – an area the size of Wales – with these reports. I told the minister that we needed concrete action, not paper. We needed safe food and water, not red tape and bureaucratic reports.

When we finally rumbled into the airport of Ust Kameno-gorsk, the capital city of the East Kazakhstan region, or oblast as it is called, a small crowd had gathered to meet us. Several journalists and a camera crew had been alerted to our arrival and word had got around. A group of teenagers had received permission from their school to come and speak to us. As Kimberley and I gave interviews to the TV crew, the teenagers listened intently. Suddenly a young Kazakh lad of around 17 or 18 stepped forward. He grasped my hand and in perfect English said, 'I want to thank you on behalf of all of the young people of Kazakhstan for what you are doing. This is our country and our future and the Polygon is our inheritance. The Polygon was constructed using the combined resources of the 15 republics of the former USSR. Now only one republic – our country, Kazakhstan – has been left to tidy up the mess. We have a lot of work to do and we appreciate your help.'

For Kimberley and me to hear such eloquent sentiments from such a young person was deeply moving.

The UN Secretary General Ban Ki Moon visited Kurchatov, the military city at the centre of the Polygon, and Ground Zero on 6 April 2010. In my role as 'Roving Ambassador' for the OSCE, I had been invited to join him. In Kurchatov, we toured the Polygon museum, where we were able to sit behind the actual control panel from which the first bombs were detonated. We viewed archive film footage of the early blasts and gruesome shots of tethered livestock flinching in horror milliseconds before being vaporised.

Ban Ki Moon chose the occasion of his visit to call for global nuclear disarmament. I accompanied the Secretary General on his visit, which involved the deployment of no fewer than five helicopters crammed with officials and journalists. The Kazakhs had laid on a special press conference at Ground Zero, although it seemed to me that the combination of helicopters and the inevitable dust storm they create on landing was not perhaps the cleverest way to show the Polygon to Ban Ki Moon. This mistake was compounded when the Kazakh foreign minister ceremoniously handed a piece of glassified rock to the Secretary General as a souvenir, explaining that it was a piece of earth from the test site which had been turned to glass by the nuclear explosion. A game of 'pass the parcel' ensued, with the nuclear rock being handed rapidly from person to person. I doubt very much if it has pride of place in Ban Ki Moon's New York office!

Despite all this, the Secretary General praised the example of President Nursultan Nazarbayev of Kazakhstan, who shut down the Polygon nuclear test site on 29 August 1991 and cleared the newly independent Kazakhstan of nuclear weapons. Mr Ban said he would recommend to the UN that 29 August should become Nuclear Non-Proliferation Day worldwide. With rogue states such as Iran and North Korea trying to develop their own nuclear weapons, that message is as important today as ever.

Back at Semey airport, hundreds of people had gathered to see Ban Ki Moon. Security was tight, but the Kazakh foreign minister and the local *akim* invited the UN Secretary General and his wife and me to a private room for a farewell drink. Inevitably, a boiled sheep's head materialised, and I was astonished when Ban Ki Moon deftly took the carving knife and sliced off the ears and meat from the cheeks, handing it around the assembled guests in traditional Kazakh fashion. Clearly he had done this before. He also drained his glass of vodka for each of the many toasts, without so much as the blink of an eye. You could see the Kazakhs were impressed.

I told Mr Ban that there has been considerable investment by the Kazakh government in new hospitals and medical facilities serving the population of Semipalatinsk Oblast and the Polygon. The new oncology hospital in Semey is a state-of-the-art facility which may become recognised globally as a world-class centre for the treatment of cancer. Nevertheless, due to the high incidence of cancers in the remote villages of the steppe, there is a need for further investment in village clinics and for adequate funding for the dedicated doctors and clinicians who staff these facilities.

Kazakhstan is steadily rebuilding itself from the virtual ruins it inherited following the collapse of the Soviet Union. It has made great headway with limited resources so far, but cannot be expected to meet the burden of the Soviet nuclear legacy alone. International recognition of the plight of the population of the Polygon must be enhanced and further financial assistance sought and properly directed towards village medical facilities, schools for handicapped children, orphanages and hospices, with some funding earmarked for co-financing the running costs of these facilities.

Some people have said to me, 'If there are now only one and a half million people living in the Polygon, why not simply move them out, provide them with new homes somewhere else and close the Polygon down?' This may sound like a simple answer, but it is not a solution at all. The Polygon is a vast area. It is also an area of great natural beauty, with the rolling plains of the steppe and its vast skies and landscapes. It was the birthplace and home of Kazakhstan's most famous poet, philosopher and humanitarian – Abai Kunanbaev. Closing the Polygon and moving the people out would be like shutting down Stratford-upon-Avon, the home of Shakespeare, and relocating the local population to Scotland. It is not an option.

The Kazakhs love this area, and indeed were proud when Stalin picked Semipalatinsk as the site for his nuclear testing zone. The huge inward investment and military infrastructure brought temporary prosperity to the area. But it also brought environmental degradation, sickness and despair. One elderly villager in Kainar said to me. 'I have only ever been happy twice in my life. The first time was when I heard the news that Stalin had chosen us for his nuclear test site. The second time was when I heard the site had been closed down!'

CHAPTER FIVE

Anatoly Matushenko and the Birth of the Soviet Bomb

Intent on uncovering more about the secret history of Stalin's nuclear tests and the way he waged war on his own Soviet citizens, I asked Elena Kachkova to try to identify a knowledgeable source in Moscow. Using her extensive networks, she came up trumps and so, in the summer of 2005, we travelled to the Russian capital to meet Anatoly Matushenko. He is a renowned expert on the history of the Soviet atomic bomb policy.

Matushenko is 69 years old but looks and behaves a lot younger. He still works in MinAtom, the ministry responsible for atomic energy in Russia. Matushenko, Elena and I met at a local restaurant near the ministry at 3 p.m., as he had not had time to have lunch yet. He told us that we were the third group that day to interview him on the problems in the Semipalatinsk region of Kazakhstan. Apparently a TV crew from a popular Russian programme came to see him earlier about a 72-year-old woman, whose 'victim of radiation' status was taken away from her by authorities in the Polygon. She was now appealing to a court in Kazakhstan and resorting to help from the TV programme to get her benefits back.

Anatoly Matushenko has a shock of unruly white hair and a white moustache with ginger edges, denoting his nicotine addiction. I started by asking him about his background. He told me: 'I graduated from the Naval Military Academy as an officer in 1959. I was trained to serve on nuclear submarines, but the country did not have any then. At the same time the rocket industry was rapidly developing. In 1960, I and five of my fellow officers were posted to serve at the Semipalatinsk test site. I joined the department, studying radioactive fallout. It meant a flat and a decent food ration. I got married in 1962, and my wife joined me in 1964, when she gave birth to our son at the Kurchatov maternity hospital. My wife and I lived on the Polygon for 13 years.'

'By the way,' Matushenko said, 'Gorbachev lied to us when he promised to keep us military scientists in active military service until the retirement age of 60. When I was 54, a lot of my colleagues were retrained or forced to retire on miserable pensions, so I figured out it was time to leave the army. I applied to the Ministry of Atomic Energy, which in 2004 was downgraded to become a Federal Agency of Atomic Energy (Rosatom).'

A waiter approached and asked if we were ready to order lunch. As we did so, Matushenko explained that unlike the United States, the Soviet Union did not keep detailed historical records of the Soviet atomic project, even of the major events. A regime of strict secrecy meant that only a few of the top leaders of the project had a full picture of events as they were unfolding. Others had to be satisfied with separate fragments of the whole mosaic. He paused briefly to give the waiter his order. 'Trying to create an accurate account from these fragments is a very difficult task,' he said. He reached down under the table and fumbled with his briefcase from which he took a thick notebook which he laid on the table. He opened the notebook and glanced at the first page of detailed notes scribbled in longhand. I caught myself nervously glancing over my shoulder, certain that the KGB were about to pounce and that I would spend the rest of my days in the cells of the Lubyanka, convicted of espionage.

The waiter returned with a bottle of vodka in an ice bucket. He unscrewed the cap and poured each of us a shot. Matushenko wished us good health and happiness and downed his vodka in a single gulp. The waiter immediately refilled his glass. He continued: 'Recently, some members of the former Soviet secret service have exaggerated the role the intelligence community played in the development of Soviet nuclear weapons. We say this with all due respect for the successes, efficiency and professionalism Soviet intelligence demonstrated in acquiring materials from abroad.'

He lit a cigarette and tilted his head to the side, blowing a long stream of smoke in the direction of the ceiling. 'Many participants in the heroic days of the Soviet bomb programme are no longer with us. And after decades of forced silence, the memoirs of survivors that have begun to appear are inevitably coloured by subjective tones. They sometimes contain unintentional inaccuracies and aberrations. Reconstructing the events of the time requires particular care, responsibility and precision. Furthermore, now that important documents are becoming accessible to the public, and artificial secrecy is being abandoned, and contacts and cooperation are being established with foreign colleagues in those fields of atomic technology which were off limits before, it is important to correct some widespread misperceptions and mistakes with regard to the creation of Soviet nuclear weapons.'

Matushenko's finger traced the lines of Cyrillic script in his notebook as he talked: 'The exceptional secrecy surrounding nuclear weapons programmes, both in our country and abroad, is well known. For a long time we used "coded language", even in our scientific reports, to disguise what we were really doing from potential spies.' He glanced down at his notes again.

'We even faked the name of the labs where the main nuclear experiments were being conducted. Everyone now knows that "Laboratory No. 2 of the Academy of Sciences" was an alias for the Kurchatov Institute.'

Matushenko continued: 'Few people knew the meaning of the abbreviations for the first Soviet nuclear and hydrogen tests. The abbreviations, invented by General V.A. Makhnev, one of feared KGB chief Lavrenti Beria's assistants, were: RDS-1, RDS-2, and so on, for *Reaktivnyi dvigatel Stalina*, or "Stalin's rocket engine". Makhnev was later very proud of his invention. Many people know that the West called our first explosions "Joe 1" and "Joe 2", after Stalin. The West was close to a correct decoding.'

Matushenko turned the page in his notebook. 'In November 1959, Vasili Yemelianov was one of a small group of Soviet experts who visited the United States. He brought back a recently published book by a Manhattan Project scientist, Arnold Kramish. The Manhattan Project was the top-secret research and development programme led by the US, with the participation of Canada and the UK, to develop the first atomic bomb at the end of the Second World War. The book contained a description of the origins of the Soviet atomic energy programme based on information available to Americans at that time. The group wanted to translate it into Russian and publish it in our country. But that idea was soon abandoned. It was decided that if it was published in the Soviet Union, the book would indirectly confirm items that were still considered state secrets.'

Matushenko went on: 'The absence of solid information created fertile ground for fantasy. In the early days of the Soviet atomic programme, the United States was the sole possessor of the atomic bomb. But a rumour spread among the population, as if to create a balance, that the Soviet Union had a new weapon of its own, with tremendous destructive force. This one, though, was a low-temperature weapon, a bomb that would instantly freeze everything around it. Even some scientists told versions of this story. As late as 1950, the Vice-rector of Moscow University, G. D. Vovchenko, told students during a chemistry lecture that "a hydrogen bomb is when you pour liquid hydrogen on the ground and freeze it completely".

'In the first years at Arzamas-16 not everyone realised that they were actually engaged in nuclear weapons work. One of the key scientists in the early development of the Soviet nuclear bomb, Yevgeni Negin, recalled that one of the chiefs of the design bureau was talking to his colleagues on the eve of the first test of the Soviet hydrogen bomb, leaning against the very device. "It is incredible how far secrecy goes in our country!" he said. "Somewhere there is another centre where they also work on weapons, and we don't even know about it! Yesterday Malenkov delivered a speech and said that hydrogen weapons have been created in our country. And we don't even know who did it and where!" This was in August 1953.'

Matushenko paused as the waiter approached our table carrying three plates of salmon eggs and freshly buttered bread. After spooning a liberal quantity of the bright red fish eggs onto a thick slice of buttered bread, he took a bite then continued to speak through mouthfuls of the succulent delicacy.

'Our Tu-4 aircraft was a copy of the American B-29. On the basis of the dimensions of the bomb-hatch, the external diameter of the bomb should not exceed 1,500 millimetres, and the length, 3,325 millimetres. By subtracting the thickness of the ballistic case of the aviation bomb and the thickness of the case for the spherical charge, which ensures that the design will be strong enough, we will obtain the initial dimensions of the spherical charge of the high explosives. This will define the dimensions of all the design elements in that spherical charge.'

Matushenko wiped his mouth with a napkin. 'The reality was very different. When work on the first bomb was under way, Yuli Khariton, who was in charge of the Soviet nuclear test programme for 45 years at Kurchatov, went to the Tupolev design bureau, the leading Moscow-based aircraft and defence centre, to meet with aircraft designers to make sure that the new bomb would fit into the bomb-hatch of the Tu-4 and to discuss other aspects of transporting the bomb by air.

'As is now well known, the design of the first Soviet atomic bomb was based on a rather detailed diagram and description of the first American bomb, which the Soviet Union obtained through the efforts of Klaus Fuchs – the notorious German–British spy who worked on the Manhattan Project and Soviet intelligence. Soviet scientists received these materials in the second half of 1945. After the experts from Arzamas-16 confirmed that the information was reliable (which required a great number of meticulous experiments and calculations), a decision was made to use the American design, which was tested and known to work, for the first explosion. Given the tension between the Soviet Union and the United States at the time, and the scientists' need to achieve a successful first test, any other decision would have been unacceptable and simply frivolous. The decision to use the American design, and information about how it was obtained, were strictly secret.'

Matushenko poured some more of the ice-cold vodka into our three shot glasses, handed one to me and one to Elena, and raising his own glass said, 'Here is a toast to our nuclear history and to a world free from nuclear bombs.' He threw back the vodka in one gulp and continued with his narrative: 'After Klaus Fuchs's trial in early 1950, it was common knowledge in the West that he had worked for the Soviet Union, but for us Fuchs's involvement remained a secret. Moreover, this secret was "sanctified" at the highest level. On 8 March 1950, Tass, the main news agency for the Soviet Union, released a special statement: "The Reuters agency has reported the recent trial in London of the English atomic scientist Fuchs, who was sentenced to 14 years in prison for divulging state secrets. British Attorney-General Sir Hartley Shawcross stated during the trial that Fuchs passed atomic secrets to 'agents of the Soviet government'. Tass is authorised to state that this statement is a crude fabrication because Fuchs is unknown to the Soviet government and no 'agent' of the Soviet government had any contact with him."

'The shock felt by surviving atomic scientists when they learned that the first device was a copy of an American bomb is understandable. They had believed the device, or more precisely its design, to have been an achievement of Soviet scientists and designers until very recently. But revealing the truth about the first test by no means lessens the significance of these pioneers' achievements.'

I could understand Matushenko's viewpoint. At that dramatic moment in history, when the threat of atomic attack hung over the Soviet Union and millions of human lives were at stake, it was the only logical decision that the Kremlin believed it could take. In addition, in order to build a real device from the American design, it was first necessary to create an atomic industry and to train highly qualified people. All of this had to be achieved in a country devastated by war. After all, did the

Americans have any doubts or reservations about their actions when, fearing that they might be overtaken by Nazi Germany, they enlisted the greatest physicists in the world in what became, in effect, an international project to create the atomic bomb?

After deciding to use the American design for the first atomic detonation, Soviet scientists temporarily slowed down the development of their own, original and more effective design. Nevertheless, their experimental work resumed in the spring of 1948, and in 1949 a detailed report provided the experimental basis and calculations for a new design that, without a doubt, was more advanced than the American design. This device was successfully tested in 1951. Its explosion was the second test of an atomic weapon in the Soviet Union.

Today, in the nuclear weapons museum at Arzamas-16, models of both devices, the one based on the American design and the design tested in 1951, are exhibited side by side. The bomb based on the Soviet design weighs half as much as the copy of the American one, but it was twice as powerful. And the diameter of the new bomb was significantly smaller.

Matushenko signalled to the waiter and ordered three coffees. He continued: 'Recently, a number of stories have appeared in the press purporting to describe how atomic weapons were created in the Soviet Union or explaining why they were not created earlier. Both journalists and mere storytellers have claimed that some far-reaching discoveries on the path to atomic weapons were made before the war, but their significance was underestimated. Such tales are, of course, pure nonsense.

'In another preposterous story, it is said that the Americans dropped not two, but three, atomic bombs on Japan. One did not explode, but was preserved. This third bomb was supposedly handed over to the Soviet Union by the Japanese.'

I said to Matushenko: 'All the historic documents I have researched demonstrate that two elements were critically important to the successful development of Soviet atomic weapons. The first was that the Soviet atomic project received the highest priority as a matter of national security. But the second was a group of Soviet physicists who had made outstanding advances before the war.'

'You are absolutely correct,' Matushenko said. 'These brilliant young scientists, led by the young Igor Kurchatov himself, had managed to reach the most advanced levels of world science and to do research of outstanding, pioneering significance.'

I waved to the waiter, indicating that we wanted the bill. Matushenko said: 'In 1943, Kurchatov wrote a famous memorandum that became a kind of manual of nuclear physics for the senior administrators of the atomic project. He pointed out that in June 1941, when research on uranium was suspended in the Soviet Union, Soviet physicists were already investigating specific ways to achieve chain reactions using ordinary uranium: with uranium 235; in a mixture of uranium enriched by uranium 235 and water; in a mixture of ordinary uranium and heavy water; and, finally, in a mixture of ordinary uranium and carbon.

'Nevertheless, apart from a small group of enthusiasts, most Soviet scientists believed that a technical solution to the uranium problem would not be found for 15 or 20 years. Meanwhile, the West was propelled by the fear that Nazi Germany would be the first to solve the uranium problem. Western scientists believed that this task could be solved in a short period of time.'

The waiter laid the bill on our table. Matushenko thanked me for lunch, then continued: 'By the end of 1941, intelligence sources reported the beginning of massive research on the uranium problem in Britain and in the United States. At once, the Soviet physicist Georgii Nikolaevich Flerov began to bombard Kurchatov and Sergei Kaftanov, who was charged by the State Defence Committee with responsibility for science, with a series of letters. In these letters Flerov asserted the necessity of returning immediately to the uranium problem and work on the atomic bomb. In spring 1942, he wrote directly to Stalin, emphasising that the creation of an atomic bomb would be a "genuine revolution . . . in military technology".

'In December 1941, Flerov sent Kurchatov a manuscript. In it he proposed using the "gun" method, that is, the rapid convergence of two hemispheres of uranium 235, to achieve a nuclear explosion. He also proposed "compression of the active material". Kurchatov never parted with Flerov's manuscript. After his death it was found in his house, in the drawer of the writing desk in his study.

'In the spring of 1942, Beria sent a memorandum to Stalin telling him about the beginning of research on atomic weapons in the West. On 11 February 1943, the State Defence Committee established a scientific and technical research programme on the use of atomic energy. Kurchatov was named as its leader.

'Two documents from early 1943 have now been declassified and are publicly available,' Matushenko continued. 'These are copies of Kurchatov's handwritten letters to the deputy chairman of the Council of People's Commissars of the Soviet Union, Mikhail Pervukhin. The first letter, dated 7 March 1943, is 14 pages long; the second, dated 22 March 1943, is 8 pages. In these letters Kurchatov compared the results of Soviet physicists and the information acquired by Soviet intelligence. He then outlined those areas of atomic research that he regarded as most promising.'

Matushenko leaned back in his chair and ran a hand through his long white hair. 'Kurchatov was an exceptional leader who organised a strategically correct programme from the very beginning. From its first day and its first steps, the Soviet atomic project had as its foundation the remarkable research of Soviet physicists, and Kurchatov devised an absolutely correct programme of implementation. But until 1945 this programme was carried out by only a few researchers who had scarce resources. The project gained real momentum only after the first American atomic explosions. It was precisely at that time that the Soviet atomic industry and technology could be developed on a broad footing, with large installations and industrial combines.

'It was exciting and extremely intense work, executed with great selflessness and enthusiasm. Kurchatov himself was the incarnation of the patriotic impulse in this unprecedented and responsible endeavour. But for some reason this heroic enthusiasm is not always noticed and taken into account by contemporary researchers, who have tended to emphasise the more unfortunate events and episodes that will happen in any large group.'

I said to Matushenko: 'In a recent history of the Soviet atomic project published in Germany, the author claims that Arzamas-16 had an atmosphere of "gallows humour". He alleges that the workers were prone to extreme cynicism and alcoholism.' Matushenko laughed. 'The author has drawn a mere caricature,' he proclaimed. 'I can assure you that, in fact, everyone was awed by the immensity of their task and fully aware of their duty.'

Matushenko claimed that there were no exceptional events that determined the atmosphere in the group, not even the peculiarities of the isolated life in a strict security zone. 'Of course there was little joy in watching the columns of prisoners who built the installation initially,' he said. 'But all that receded into the background and people had little regard for the difficulties of everyday life. They were trying to achieve success in the best and quickest way. They knew that the country was in danger, and they understood that the state was relying on them and giving them everything necessary for their work and their daily life. And they carried out their task splendidly.'

We rose to leave, and as we made our way out of the restaurant Matushenko said to me, 'As we know, historical truth can be distorted not only by subjective accounts, but by excessive secrecy, which limits access to information, or by simple misunderstanding. Conjecture also abounds when the truth is hushed up for political reasons, as in the case of Lavrenti Beria. If there is no truth today, there will be myths tomorrow. It is not my intention to reassess the crimes of this terrible man who caused immeasurable suffering. But until mid-1953, for about eight years, Beria had overall responsibility for all the work on the atomic project. In the interests of history, the facts of his administration should be known in more detail.'

Lavrenti Pavlovich Beria was chief of the Soviet security and secret police apparatus (NKVD) under Joseph Stalin. He was the longest lived and most influential of Stalin's secret police chiefs, wielding his most substantial influence during and shortly after the Second World War. He simultaneously administered vast sections of the Soviet state and served as de facto Marshal of the Soviet Union in command of the dreaded NKVD field units. These were responsible for anti-partisan reprisal operations on both friendly and enemy civilian populations, and the apprehension and summary execution of thousands of 'turncoats, deserters, cowards and suspected malingerers'. Beria oversaw the vast expansion of the Gulag slave labour camps, and was primarily responsible for the Katyn massacre of Poland's officer class and intelligentsia. He was widely feared, even by members of the Politburo, for his ruthlessness and violence.

I was fascinated by this mention of Beria, and informed Matushenko that Elena Kachkova and I had already dug around in the Soviet military archives in Moscow. There we had discovered that at first the Soviet atomic project was administered by Vyacheslav Molotov. His leadership style and, correspondingly, its results were not terribly effective. Kurchatov did not hide his dissatisfaction. Although some members of the Special Committee and the Technical Council on the Atomic Bomb complained about Beria's methods in a letter to Stalin, once the project passed into Beria's hands, the situation changed completely.

Beria understood the necessary scope and dynamics of the research. This man, who was the personification of evil in modern Russian history, also possessed great energy and a capacity for work. The scientists who met him could not fail to recognise his intelligence, his willpower and his purposefulness. They found him a first-class administrator who could carry a job through to completion. It may be paradoxical, but Beria, who often displayed great brutishness, could also be courteous, tactful and straightforward when circumstances demanded it. Matushenko said: 'Beria's meetings were conducted in a businesslike and fruitful way. They never lasted too long. And he was a master at finding unexpected and unconventional solutions.'

We were pushing our way through crowds of pedestrians as we headed for the MinAtom building, in the heat of an August afternoon in Moscow. Matushenko glanced towards me. 'Another of the pioneering atomic scientists, Mikhail Sadovskii, had quite a difficult meeting with Beria. Beria met about 30 people in his Kremlin office to discuss the preparation of the test site in the Polygon for the first thermonuclear explosion. The scientists who had been ordered to give the reports were trying to say where the equipment should be located, what kind of structures should be erected, and how and what kind of experimental animals should be placed at the site to study the impact of the blast effects. Suddenly, Beria became incensed. He interrupted angrily, moving from one person to another, asking strange questions which were not easy to answer. Finally, Beria completely lost his temper. He screamed: "I will tell you myself exactly what is needed!" It gradually became clear from his stormy monologue that he wanted everything at the test site to be totally destroyed in order to provide the maximum terror.'

We had paused at an intersection. A policeman was standing in the centre of the crossroads directing the busy Moscow traffic with a blue light in one hand. Matushenko turned to face me: 'After the meeting, the participants left in a gloomy mood. Sadovskii realised for the first time that dealing with Beria was no joke. Beria was a quick worker, and he visited all the installations to acquaint himself personally with the results of the project. During the first atomic explosion, he was the chairman of the state commission. Despite his exceptional position in the party and the government, Beria made time for personal contacts with people who interested him, even those without official distinctions or titles. On a number of occasions he met Andrei Sakharov, who was then merely a student of Physical and Mathematical Sciences, as well as with Lavrentiev, who was a recently demobilised sergeant from the Far East.

'If the project needed some specialist that Beria viewed as unacceptable, he was willing to make an exception. When the KGB wanted to remove a particular scientist on the grounds that he was unreliable, Khariton telephoned Beria directly and told him that the project needed this person. After a long pause, Beria asked a single question: "Do you need him very much?" After receiving a positive response, Beria said, "All right," and hung up. The incident was closed.'

I said to Matushenko: 'In my search of the archives, I discovered that Joseph Stalin certainly paid personal attention to the atomic project, as demonstrated by Kurchatov's notes of his one-hour meeting with Stalin in the evening of 25 January 1946, which have been preserved. The meeting was also attended by Molotov and Beria. In the course of the conversation, Stalin rejected small-scale projects and cheap solutions. He stressed that it was necessary to act "on a broad front, on a Russian scale" and that in this respect all-round help would be provided. Stalin said that Soviet scientists were very modest people, and that "sometimes they do not notice that they live badly".'

I continued: 'According to the archives, Kurchatov apparently reported how in relation to scientists, Stalin was concerned that they should improve the material conditions of their lives and that they should be given prizes for major achievements. Shortly before the first explosion of the atomic bomb, the directors of the basic areas of research concerning the preparations for the test reported to Stalin himself, with Beria and Kurchatov in attendance. The specialists were invited to Stalin's office one by one, and Stalin attentively listened to each. The first report was delivered by Kurchatov, followed by Khariton and the others.

'This was Khariton's only meeting with Stalin. Stalin asked Khariton: "Couldn't two less powerful bombs be made from the plutonium that is available, so that one bomb could remain in reserve?" Khariton, who knew that only the precise amount of plutonium required for the American-designed weapon was available, and that taking any additional risk would be unacceptable, responded negatively.'

I told Matushenko that I had read somewhere that early in October 1949, after the explosion of the first bomb, Kurchatov and other members of the commission reported the results to

Stalin. Stalin was interested in details and he asked the speakers excitedly several times whether they had personally seen what they were reporting.

On 29 October 1949, two months after the explosion of the atomic bomb, Stalin signed a secret decree that was issued by the Council of Ministers of the Soviet Union. Few people, other than those who received awards, knew of its contents and the award winners were informed only through separate, personal excerpts, so that they would not see the entire document. In this decree, several particularly distinguished participants in the research, led by Kurchatov, were named 'Hero of Socialist Labour', awarded bonuses and given ZIS-110 or Podeda cars, the title of the Stalin Prize Laureate of the First Degree, and dachas. Their children were to receive a free education in any educational establishment at state expense. The recipients and their wives were awarded free and unlimited transport by rail, air and water, anywhere in the Soviet Union for as long as they lived, a privilege shared by their children. The right to free travel was later abolished by Nikita Khrushchev.

Veteran scientists say that Beria, when deciding which prizes to award, used a simple principle that was not without a certain malicious humour. Those who in case of failure would have been shot were to receive the title of Hero. Those who would have been given life imprisonment were to be awarded the Order of Lenin, and so on down the list.

As we neared the great doors of the ministry building, still displaying the hammer and sickle of the USSR, Matushenko strode past two military guards and pushed open the door, beckoning Elena and me to follow him into the huge marble foyer. The sudden change in temperature as we entered the air-conditioned foyer, together with the lunchtime vodka, brought a pink glow to Matushenko's face.

He signalled to us to sit down on a large bench, explaining that it would not be possible to get us past security to continue our conversation in his office because we did not have the appropriate passes. He went on: 'The burden of responsibility that rested on the shoulders of the creators of Soviet nuclear weapons was no laughing matter. The scientists all felt the prickly chill of possible retribution when a device "refused" to work, as the bomb makers say, and the nuclear explosion did not take place. The first such "refusal" took place, fortunately, on 19 October 1954, after our country already possessed atomic and thermonuclear weapons, and Beria was no longer alive.

'Kurchatov, Khariton and other participants visited the site of the failed atomic explosion in a bunker in the Polygon near Semipalatinsk, where they began calmly trying to understand the causes of the failure. Suddenly, a colonel from the KGB state security appeared. Spick and span, with a service cap on his head, he saluted and addressed Vyacheslav Malyshev, our minister:

' "Comrade Minister, if I understand correctly, there was a failure?"

' "That is correct."

' "Please allow me to start an investigation."

'We all felt uncomfortable. Malyshev calmly began to speak:

' "You see, we are dealing with science here. Not war. These are new things, we don't know everything yet. The scientists are getting to the bottom of it. They also cannot say right now what the cause was."

' "So allow me to start an investigation!"

'The colour of Malyshev's face started to change slowly. He turned red. "I tell you, this is an experimental thing, made for the first time. Apparently, we were unlucky in some way and we failed. But I think that we will sort it out in a short time and find the answer."

' "So allow me to start an investigation!"

'Malyshev turned purple with rage, then screamed:

' "Get out!"

'The KGB colonel saluted again, turned on his heels and went away.'

'Today, many people realise that it was Soviet physicists who first developed thermonuclear weapons,' Matushenko said. 'We began by trying to catch up with our American colleagues, who created the atomic bomb in mid-1945. By August 1949, the dangerous US monopoly on atomic weapons was eliminated. Then, Soviet physicists took the lead, and on 12 August 1953 they were the first in the world to explode a real hydrogen charge, which was ready to be used as a bomb. The idea for that device belonged to Andrei Sakharov.'

I pointed out to Matushenko that American scientists carried out a thermonuclear explosion on 1 November 1952, but, he replied, 'This experiment was only a step on the way to the creation of a true hydrogen bomb.

'The US device was a huge, immobile 50-tonne land-based thing the size of a two-storey house,' he said. 'The nuclear fuel contained inside it had to be cryogenically condensed. Soviet scientists managed without this kind of very complicated and costly experiment. The United States took the lead in improving hydrogen weapons in 1954. But by 1955, Soviet physicists had made a genuine technological breakthrough, neutralising the Americans' success. Moreover, Soviet scientists were the first in the world to explode a hydrogen bomb dropped from an aeroplane. This experiment took place on 22 November 1955.'

Matushenko went on: 'On 30 October 1961, Soviet physicists detonated a 50-megaton bomb, which remains unsurpassed in terms of its yield. This device was distinguished by its purity. Ninety-seven per cent of its energy yield was derived from thermonuclear reactions. The complete success of this test proved that it was possible to design devices of virtually unlimited power on the basis of the principle proposed by Sakharov. The bomb was exploded at an altitude of four kilometres over Novaya Zemlya, using a Tu-95 strategic bomber.'

Matushenko said: 'Looking back, we know that one of the initial impulses for both the US and Soviet atomic projects was the fear that Nazi Germany, which before the war possessed the most advanced and modern technologies and first-class science, might create atomic weapons.'

'Hitler's declarations about weapons of retribution sounded sinister. After the Second World War, nuclear weapons became the basic argument in the extremely dangerous confrontation between the world powers. As we gradually free ourselves from the legacy of confrontation, we can now see that in the late 1940s and early 1950s nuclear physicists, including the brilliant constellation of Soviet physicists headed by Igor Kurchatov, did something greater and more lasting. They opened the door to a new era. In this era, atomic energy not only defines the technological level of society, it also influences culture, politics and the future. Consequently, it influences history.'

I said: 'It is frightening to think how different events might have been if Hitler had been able to develop an atomic bomb. The outcome of the Second World War would have been entirely different.'

As Matushenko pondered this horrifying thought, I took the chance to ask him what, in his view, the people of Semipalatinsk are suffering from. He replied: 'The illness that people are suffering from in Semipalatinsk has a name – it is called radiophobia. On top of that, people were enticed to go and settle down there with promises of government subsidies in the form of housing, food and travel allowances, and now those subsidies have been taken away. Rather than improving social conditions in order to stop people complaining, the government of Kazakhstan took away whatever little people had. The people feel cheated. They are also suffering from radiophobia, and they consider radiation their biggest enemy.'

He continued: 'I could have considered myself to be a victim of radiation too. I took part in all the tests, starting with the one in 1961. Actually I was present when the last atmospheric and above-ground explosions were conducted in 1961-2. Then they were stopped in 1962, after which the tests continued underground. The name of my department was the Department for Radioactive Contamination Research, and my work included exploration of the irradiated territory, collection and analysis of samples, visits to local villages, taking measurements there, installation of specialised equipment – so you can tell I was right in the epicentre of it all.

'And I am not alone. I have a lot of colleagues and friends who have been there, and none of them is panicking, because they are not susceptible to radiophobia. You must realise – we were in the centre of things, metres away from "danger", but

I get complaints from people who were tens and even hundreds of kilometres away. But when a person starts seeing an enemy in radiation, despite the fact that we say: "You must not be afraid of radiation, but you must treat it with respect", when people lose this respect and succumb to blind panic, then radiophobia takes its toll. Many people have lost their lives because of this. Have you heard of illnesses brought on by emotional stress? When you get a spot, or a headache, and assume that radiation is to blame, when in reality it is due to excessive levels of molecular zinc in water, or manganese, or some other trace element that makes your hair fall out, for example, but nobody bothers to check because the blame is already on radiation.

'Plus there is also the desire to get more social protection, benefits – to take money out of someone else's pocket. There are fights going on, no nerves are spared, it is very draining emotionally and physically. In reality, people were given certain benefits by law, then time passes and a new policy is adopted, diminishing or completely scrapping these benefits. And you know our bureaucrats, all they say is "Go to court, the courts will decide." But who will go there? I won't, my wife won't go either. But this woman I was telling you about earlier, she is 72, and she is so determined to win her court case, even with the help of that television programme. What will probably happen is that she will lose her case anyway and be crushed by the system, and this will make her health even worse.'

Matushenko extended both his arms towards me, palms upward. 'We are saying, President Nazarbayev does not need a scared population infected by the virus of radiophobia, he does not need a nation that is ill, complaining all the time and only managing to survive on benefits. The Polygon is a unique laboratory. There is a unique chance, not just for Kazakh scientists, but for experts from around the whole world, to study the conditions created at the site, to study the behavioural or other problems of the local population and to undertake joint projects to train new radiation inspectors and auditors.'

'The locals who demanded the closure of the Polygon are now suffering the social consequences of it after the funding from hospitals was withdrawn and the infrastructure decommissioned. They can now see that more harm was done by the closure. The social infrastructure was destroyed, the collective farms disintegrated. Radiation is not the enemy; the collapse of the social structure became the worst enemy.

'And it is beneficial for certain individuals or political forces to conceal real facts and radiation contamination data in order to gain popular support to achieve their own political ambitions. As for the land surrounding the test sites in the Polygon, it was perfectly suitable for agriculture and pastures,' said Matushenko. 'There are, of course, so-called "red" zones, but the work of re-zoning the Polygon for agricultural purposes will be carried on later this year. There are three districts, mainly adjacent to the Semipalatinsk and the Pavlodar Regions, which have to be fenced off. At the moment anybody can go there with their cattle, and people do, because nobody has taken responsibility for those lands. They also come to dig up copper cables and steal remaining equipment. We will help the National Nuclear Centre to gain control of those zones to create ecological laboratories to study the environment there.

'And as for the radiation damage to the health of the population of adjacent villages, we published a book on the contemporary radio-ecological status of test sites in 2005. I will give you a copy. This is the most comprehensive book to date based on research conducted by the Russian, Kazakh and American scientists during the last ten years, with the help of the Russian Ministry of Health, and MinAtom.'

Matushenko held out his hand to me. 'I must return to my office,' he said. 'I hope you have found our discussion useful?'

'Can I ask you one last question,' I countered. 'What is the correct number of detonations conducted in the Polygon, as I have encountered a lot of conflicting figures, anything from 400 to well over 600?'

Matushenko replied: 'There are official tables documenting all the explosions conducted in the Soviet Union, the total of which is 715. Of these, 124 were "peaceful" detonations for industrial purposes, so it is not right to claim that more than

600 took place in the Semipalatinsk Polygon, as there were other Polygons, like Novaya Zemlya. The correct figure for Semipalatinsk should be around 450, if my memory serves me right.'

I shook Anatoly Matushenko's hand and thanked him warmly for the amount of time he had devoted to our interview and the huge well of information he had imparted. He stood and waved goodbye as Elena and I headed back out into the fetid Moscow streets.

The Other Polygon – Novaya Zemlya

As Anatoly Matushenko had pointed out, Central Asia was not the only part of the former Soviet Union to suffer the consequences of Stalin's nuclear ambitions. Novaya Zemlya is an archipelago situated in the Arctic Ocean at the very extreme north-east of Europe and the north of Russia. It consists of two main islands, Severny (northern) and Yuzhny (southern), which are separated by the Matochkin Strait. Novaya Zemlya itself separates the Barents Sea from the Kara Sea.

Novaya Zemlya is a mountainous area that is rich in natural minerals such as copper, lead and zinc. The northern island consists of many glaciers, while the southern island is characterised by tundra landscape. The ecology of the region is influenced by the extreme climate, but Novaya Zemlya nevertheless supports a diverse ecosystem. One of the most notable species present is the polar bear. In 2002, the indigenous population of 2,716 mostly resided in the administrative centre of Belushya Guba. They are called the Nenets, and survive mainly by fishing, trapping and hunting polar bears and seals.

Russian hunters first visited Novaya Zemlya in the eleventh century but it was not until the search for the north-east passage began in the sixteenth century that western Europeans became aware of the area. The first permanent settlement was established in 1870, when small numbers of Nenets people were resettled there in a bid by Russia to keep out the Norwegians. This population, then numbering about 1,500, was removed in the 1950s, when Novaya Zemlya became notorious as a nuclear testing area.

Novaya Zemlya was first designated as a nuclear testing site in July 1954. During the Cold War – as in the Polygon – hundreds of nuclear tests were conducted in the region, including the explosion of the legendary Tsar Bomba on 30 October 1961. This was the largest, most powerful nuclear weapon ever detonated. In September 1961, two propelled thermonuclear warheads were successfully launched to target areas on Novaya Zemlya. Subsequently, rockets of this type were deployed to Cuba, causing the Kennedy/Krushchev nuclear stand-off and almost sparking an atomic war.

In 1963, the Limited Test-Ban Treaty was implemented and most atmospheric nuclear tests were outlawed. Consequently, the Soviets turned to underground testing, and in 1973 four nuclear devices of 4.2 megatons total yield were detonated on Novaya Zemlya. Although far smaller in blast power than the Tsar Bomba and other atmospheric tests, the confinement of the blasts underground led to tremors rivalling natural earthquakes. In September 1973, a nuclear test triggered seismic activity of a magnitude of 6.97 on the Richter scale and set off an 80 million tonne avalanche that blocked two glacial streams and created a lake 2km in length, rather like the Atomic Lake in Semipalatinsk.

Over its history as a nuclear test site, Novaya Zemlya hosted 224 nuclear detonations, with a total explosive energy equivalent to 265 megatons of TNT. In comparison, all the explosives used in the Second World War, including the detonations of the two US nuclear bombs in Hiroshima and Nagasaki, amounted to only two megatons. In 1988–9, the Novaya Zemlya testing activities increasingly became public knowledge, and in 1990 Greenpeace activists staged a protest at the site. The last nuclear weapons test ever conducted in the Soviet Union and Russia took place on Novaya Zemlya in 1990.

Tsar Bomba was constructed by the Soviet Union under direction from Nikita Khrushchev. The original design was for a 100-megaton weapon. By comparison, the largest US detonation was the 15-megaton Bravo blast. But in the device that was detonated, some of the uranium was replaced with lead. This had the effect of also reducing fallout. If the full yield design were detonated, the global fallout impact would have been significant. The bomb itself weighed 27 tonnes and, consequently, had limited military use as a readily deployable weapon.

The Tsar Bomba was released at over 30,000 feet and dropped by parachute to 13,000 feet, where it detonated on 30 October 1961. The rationale for detonating this device was largely for propaganda purposes. Khrushchev announced the testing ahead of time, putting great pressure on the scientists responsible for the assembly and detonation. In 1963, Khrushchev claimed the device had been weaponised and was located in East Germany. However, no normal aircraft capable of handling it had ever been observed in Europe, and it was generally believed that this was a bluff.

Due to the nuclear testing conducted on Novaya Zemlya, there has been serious pollution of the Kara and Barents Sea, as well as contamination of the water surrounding Novaya Zemlya, the land itself and parts of Europe, particularly Norway. On 2 August 1987, the Soviet Union exploded a bomb on the island of Novaya Zemlya that was so large, radioactive fallout was measured all over Europe.

Norway is particularly affected by the legacy of testing at Novaya Zemlya, given the fact that it lies only 900 kilometres from the islands. The Barents Sea contains much of Norway's rich fishing grounds, which have been severely affected by radioactive fallout following 36 years of nuclear tests. Of the 42 underground explosions at Novaya Zemlya, 25 were accompanied by the release of radioactive inert gases. There were three underwater explosions, each less than 20 kilotons, but most of the radionuclides remained in the water and sediments. A total of 17 reactors were dumped in the Barents Sea, to the west of Novaya Zemlya, including seven containing spent nuclear fuel.

In 2004, the Russian authorities admitted an increase in cancers and other illnesses among the indigenous population, similar to the type of diseases observed in the Semipalatinsk

Polygon. The local island people contract liver cancer at a rate ten times the national average, and the death rate from cancer of the oesophagus is considered amongst the highest in the world. There is also a high incidence of premature deaths from cancer, and of infants born underdeveloped. Novaya Zemlya was not only used for testing nuclear weapons, it has also been used as a dump for radioactive waste and a graveyard for nuclear arms, submarines and reactors which have become obsolete. This accumulation of dangerous radioactive material has exacerbated local health problems.

CHAPTER SIX

Uranium Tailings

Stalin's nuclear weapons programme could not have succeeded without access to a plentiful supply of uranium. The countries of Central Asia were the main suppliers of uranium for the Soviet military machine. Four of the five Central Asian republics were involved in this trade: Kazakhstan, Kyrgyzstan, Tajikistan and Uzbekistan. In fact, the first nuclear bomb detonated in the Polygon used uranium from Tajikistan.

Commercial mining and milling of uranium ore produces large amounts of radioactive waste. Two types of waste are produced: one comprises solid radioactive waste from low-grade unusable ores stored in dumps, and the other consists of solid, liquid and gaseous radioactive and chemical wastes from the plants producing uranium oxide. The latter has to be stored in large reservoirs called tailings impoundments.

The final product of the mining process, uranium oxide, or 'yellow cake', is only slightly radioactive because approximately 70 per cent of the radioactivity is left behind in the tailings. Tailings are problematic if they are allowed to dry out, as they will emit radon gas into the air, and radioactive particles can then be picked up and transported by the wind. To prevent such contamination, tailings are normally submerged underwater.

While underwater storage may help to minimise the problem of radioactive particles being blown about, it will increase the likelihood of groundwater contamination. To offset the risk, the tailings dumps, which were all constructed during Soviet times, were lined with plastic liners. But these plastic-lined ponds were constructed decades ago, and there has been little or no maintenance since. Now the plastic has become brittle, and in many ponds it has ruptured altogether, causing groundwater contamination. At many of these former uranium processing plants fences are poorly maintained, and cattle and other livestock often graze around the dumps.

Unbelievably, there is more than 812 million tonnes of radioactive waste in the tailings dumps of still active and former uranium mines in Central Asia. This is a staggering amount, and creates an enormous problem that is hard to imagine, never mind resolve. The tailings are highly susceptible to natural disasters and represent a significant environmental hazard due to their close proximity to the main river systems of Central Asia. They pose a serious ecological threat on a regional scale.

As I was repeatedly told during my frequent visits to the Central Asian republics, large amounts of radioactive waste stored in these conditions can not only cause contamination of water resources, but there is also a growing risk of radioactive waste storage sites being destroyed by natural disasters such as earthquakes or floods, or even man-made catastrophes. The most dangerous radioactive waste storage sites in Central Asia are concentrated in the so-called 'Ferghana radioactive belt', which is home to over 10 million people in Kyrgyzstan, Tajikistan and Uzbekistan. The Ferghana Valley is one of the poorest areas on earth, and a seething hotbed for Islamic fundamentalism. Environmental problems in the Ferghana Valley could spark spontaneous outbursts of violence and conflict. Uranium tailings hot spots which are areas of serious concern include Mailuu-Suu, Min-Kush and Kaji-Say in Kyrgyzstan; Charkesar in Uzbekistan; and Taboshar and Degmay in Tajikistan.

Kazakhstan has about 20 per cent of world uranium resources, with 529 radioactive waste storage and dump sites. The total volume of waste in the tailings of the former uranium industry in Tajikistan is around 55 million tonnes. Kyrgyzstan is one of the primary areas for uranium extraction, and there are 29 uranium tailings dumps containing more than 41 million cubic metres of radioactive waste in the country.

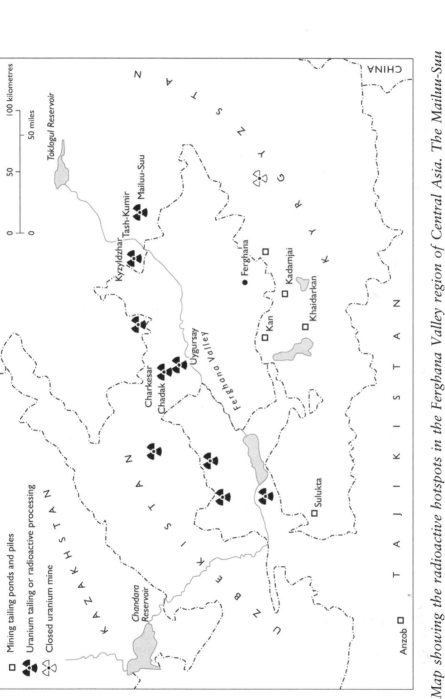

Mining tailing ponds and piles

Uranium tailing or radioactive processing

Closed uranium mine

Chandara
Reservoir

KAZAKHSTAN

UZBEKISTAN

Charkesar

Chadak

Uygursay

Ferghana Valley

Anzob

Sulukta

Kan

Ferghana

Kadamjai

Khaidarkan

KYRGYZSTAN

TAJIKISTAN

CHINA

Kyzyldzhar
Tash-Kumir
Mailuu-Suu

Toklogul Reservoir

0   50   100 kilometres

0   50 miles

*Map showing the radioactive hotspots in the Ferghana Valley region of Central Asia. The Mailuu-Suu
uranium mining plant can be seen in Kyrgyzstan and the Taboshar plant in Tajikistan.*

Source: http://maps.grida.no/go/graphic/radioactive_waste_hotspots_and_transboundary_pollution_in_central_asia_s_ferghana_valley

One of the largest extraction sites is the Mailuu-Suu plant, and in the surrounding area the number of people with cancer is almost four times higher than average for the country as a whole. Also, the number of children born with congenital diseases is 2.8 times higher, again attributed to pollution from the uranium tailings dumps.

Uranium Mining in Kyrgyzstan

During the Soviet regime, large uranium deposits were discovered and developed north of the Ferghana Valley in what is modern-day Kyrgyzstan. The uranium mined on Kyrgyz territory was utilised by Soviet scientists to enhance their nuclear capabilities and develop warheads, although nuclear weapons were not produced or deployed on the territory of Kyrgyzstan. The current problem is that the waste products left by the uranium industry, especially the uranium mining facilities, now present a challenge, not just to the environment, but also to public health and domestic political stability in Kyrgyzstan, and even to regional, political and economic stability.

Mailuu-Suu is a particularly problematic plant, not only because of the huge volume of waste stored there but also because it is located in a highly seismic zone, within which earthquakes with a magnitude of 7.0 on the Richter scale are a regular occurrence. Such strong earthquakes can trigger major landslides, which in turn could destroy tailings dumps or block the river. The Mailuu-Suu site is a prime example of how an established environmental problem can evolve into a threat jeopardising both domestic and regional security.

The best example of this is the natural landslide that occurred in May 2002. A massive landslide was caused by six weeks of torrential rain in the south of the country. This eventually blocked part of the Mailuu-Suu River, whose waters then threatened to flood radioactive tailings dumps located along its banks. If the tailings dumps had been flooded, the river would have carried the radioactive waste downstream through the watershed of the Ferghana Valley. Not only would this have had serious environmental implications for the downstream regions of Kyrgyzstan, Uzbekistan and Tajikistan, but it would also have added to simmering political tensions in the region.

Another ominous dimension that is less discussed is the security threat that these sites could potentially pose. This is particularly an issue in the Central Asian region, which has witnessed political instability and incursions by Central Asian Islamist groups believed to have terrorist links. Highly radio-active materials, which could be used to produce radiological dispersal devices (RDDs or 'dirty bombs') might be present inside the tailings, as well as in abandoned equipment at these sites. Both the tailings dumps and so-called 'orphan' sources, which could contain reactor-produced isotopes, might present security risks if left unmonitored and unguarded.

Although the contents of uranium tailings themselves do not pose a large-scale RDD danger, no one seems to know for sure if other dangerous radiological materials exist at these former mining sites, due to their large-scale abandonment after the break-up of the Soviet Union and lack of access to Soviet-era documents that might contain such information.

In March 2010, I met Akyle Aytbaev, the deputy minister for emergency situations for the Republic of Kyrgyzstan in his rather dilapidated office in Bishkek. I had flown in to the international airport 16 miles from the capital in the small hours of the morning. Known as the Manas air base, this airport plays host to the US Air Force, supporting Operation Enduring Freedom and the International Security Assistance Force in Afghanistan. It was a hive of activity in the middle of the night, with great arc lights casting a yellow glow on mammoth troop transporters which were loading and unload-ing troops and *matériel*.

I was met by a team from the Kyrgyz Ministry of Foreign Affairs and driven into the capital. It was hard to know we were driving through a large city, as Kyrgyzstan was experi-encing one of its regular electricity blackouts and there were no street lights or any other indications of life, apart from the occasional office or garage with its own generator.

After a few hours' sleep, I was driven to the Emergency

Situations Ministry. On learning that I was from Scotland, the minister said he had always wanted to visit Loch Ness and see the monster. 'Uranium tailings are our "monster" in Kyrgyzstan,' he continued. 'It is one of the biggest dangers facing Central Asia. Only $6,000 has been allocated for the coming year to support the creation of a centre to tackle the problem of uranium tailings, while $44 million has been spent on seminars and international consultants to tell us what to do! The World Bank is now helping in the Mailuu-Suu case, with the rehabilitation of uranium dumps at a cost of $7 million there alone. However we need a further $28 million for total rehabilitation of all the dumps. There has been no maintenance carried out since the mines closed in 1973.'

He explained that Kyrgyzstan also has constant problems from avalanches and earthquakes. 'We have on average more than 3,500 earthquakes per year of varying strengths, and approximately 4,000 rock falls, avalanches and landslides every year,' he said. 'In 1992, we had an earthquake that measured 9 on the Richter scale and killed 53 people. Many of our people live in the mountains and valleys and are at constant risk. The military help to pre-empt dangerous avalanches by firing cannons, thus causing the snow to slip.'

Mr Aytbaev continued: 'There are 70 radioactive waste sites in Kyrgyzstan alone, including 36 uranium tailings sites. There are 92 waste dumps containing 254 million cubic metres, or 475 million tonnes, of waste.'

The minister explained that the security risks include the possible terrorist theft of radioactive materials. Although Kyrgyzstan does not possess highly enriched uranium, which can be used to produce nuclear weapons, it does harbour radioactive materials that have been abandoned in sealed or unsealed sources or are poorly secured within non-operational mining facilities. Mr Aytbaev left me under no illusions that he fears some of this material could fall into the wrong hands with dire consequences for security. Stalin's legacy could yet affect us all.

The recent history of Kyrgyzstan is replete with examples of the theft and smuggling of such material across transnational boundaries. The main obstacles in protecting the materials are lack of knowledge about the risks they pose and an inadequate inventory of potential radioactive sources, including uranium tailings sites in the Kyrgyz Republic and throughout Central Asia.

The information provided about the contents of the sites is scant. It is known that a significant amount of radioactive waste remains throughout the territory of Kyrgyzstan, although what this waste consists of is not clear. Reports indicate that the isotopes within the Mailuu-Suu site's tailings include thorium, copper, arsenic, selenium, lead, nickel, zinc, radium and uranium. These isotopes are of an environmental – although not a physical security – concern. However, it is unclear if these environmentally toxic isotopes are the extent of radioactive items that exist in Kyrgyzstan and other Central Asian territories. In addition, most of the sites have no security measures, allowing the general population to scavenge for radioactive metals and other waste. This lack of security also could result in intentional sabotage of the tailings to cause environmental degradation.

The real security problem would arise if the additional, unknown radioactive sources that may exist in Kyrgyzstan include items such as bits of mining equipment which are known to contain cobalt-60 and cesium-137 isotopes that would make attractive dirty bomb components.

The minister confirmed that the Kyrgyz government recognises the danger of unauthorised access to the radioactive waste sites and abandoned mines by scrap metal scavengers who 'open up radioactive waste storage facilities and break into abandoned uranium mines and underground workings in order to extract contaminated metal, electro-technical equipment and cables from them'.

'Various incidents of smuggling have transpired in recent years, from very minor to extremely dangerous threats warranting an International Atomic Energy Agency (IAEA) investigation,' the minister said. 'In December 2007, cesium-137 was found on a train carrying ferrous scrap metal from Kyrgyzstan. It was scheduled to be transported to Iran.

Caesium-137 is one of the most desirable ingredients for an RDD and, as such, the incident came to the attention of the IAEA, which appears still to be investigating the matter.'

There is no doubt that the threat of terrorists getting their hands on potentially lethal radiological material in Kyrgyzstan cannot be ignored. The inadequate security maintained at uranium tailing sites could present a serious danger of theft by terrorist groups, depending upon the level of radioactivity in materials that have been abandoned.

Many terrorist groups operate in and around Kyrgyzstan, including al-Qaeda. As pressure on them intensifies, particularly in the border areas of Afghanistan and Pakistan, many of these terrorists are seeking refuge in Central Asia. In July 2009, the Kyrgyz authorities arrested 18 people accused of assisting international terrorist groups. Many of those detained had been trained in Afghanistan. In early January 2011, a fierce firefight in Bishkek left four law-enforcement officers and two alleged militants dead.

Top security officials in Bishkek described the incidents as a call to arms against perceived Islamic militants. 'A war has been declared on all of us . . . We must distinguish between good and evil. Today evil is wearing the mask of a believer, trying to intimidate us, to cause panic and division,' Interior Minister Zarylbek Rysaliyev told me.

Of particular note is the terrorist group Hizb ut-Tahrir, which has grown in number by 10,000 recruits worldwide over the last two years, but which also has a significant presence in Kyrgyzstan and other countries in Central Asia. There are also indications that this group might encourage the use of weapons of mass destruction, in addition to the use of radiological materials for an RDD.

Prior to his death at the hands of the US Seals, Osama bin Laden was quoted as saying: 'It is a fact that the Islamic Republics' region is rich with significant scientific experiences in conventional and non-conventional military industries, which will have a great role in future jihad against the enemies of Islam.'

Deputy Minister Aytbaev told me that the minister of

emergency situations had made a special plea to the international community for aid for increased security at the tailings sites. 'If we do not guard the tailings dumps, then the radioactive materials can become tools in the hands of extremist groups and terrorists,' he said.

A series of high-level international conferences have been held to debate and find solutions for the problem of uranium tailings in Central Asia. Various governments, including Canada, Finland and Norway, have indicated their support for future uranium tailings projects. Whether the international community will continue to support the measures undertaken by Kyrgyzstan remains to be seen. Important strides are being made in the environmental realm, while the potential security threat is less of a focus at international conferences. However, it is clear that increased attention and financial assistance are focused on Kyrgyzstan in particular, due to its efforts in recent years to improve the situation.

Radioactive waste security in Kyrgyzstan and in other Central Asian territories is at a critical juncture, especially in view of growing Islamist links to the region. This threat seems to be acknowledged occasionally by leading Kyrgyz officials, although there appears to be little coordinated effort to address radioactive security in the context of a terrorist threat. If this is indeed a serious threat, it is likely that an inventory of radioactive materials will need to be taken and incorporated into future efforts at waste clean-up.

The instability of the region was underlined when only a week after my visit to Bishkek, on 10 April 2010, a demonstration took place in the city of Talas against government corruption and increased living expenses. The protests turned violent and spread nationwide. The Kyrgyz president Kurmanbek Bakiyev imposed a state of emergency, and the police and secret services arrested many of the opposition leaders. In response, protesters took control of the internal security headquarters and a state television channel in Bishkek, prompting President Bakiyev to flee the country. All of the ministers I had met, including Deputy Minister Aytbaev, were arrested or dismissed.

Uranium Tailings in Tajikistan

Neighbouring Tajikistan did not escape from Stalin's war on nature. The Tajiks are deeply concerned over the rehabilitation and safety of uranium tailings dumps.

In September 2010, I visited the capital, Dushanbe, for talks with Hamrokhon Zarifi, the foreign minister, and various other senior government ministers. Mr Zarifi told me that there were a large number of tailings dumps in northern Tajikistan. 'In the Soviet era, the northern city of Chkalovsk was one of the key centres of the uranium industry. Chkalovsk's enterprises processed not only Tajik uranium ore, but also uranium ore delivered from Uzbekistan, Kyrgyzstan and Kazakhstan. Radioactive waste was stored in tailings dumps that do not meet appropriate safety rules and are situated right next to residential areas and rivers. These tailings dumps pose an immediate threat to the environment.

'In Gazion, a village in northern Tajikistan, residents live and work in close proximity to an old uranium mine. The site, just 2km from the village, is one of eight in the region where uranium ore waste has been left where it was dumped when the Chkalovsk plant was processing uranium for the entire Soviet nuclear industry.

'These days, villagers admit that they graze their cattle close to the site, even though it is not allowed. They blame radiation for health problems and lower crops yields. Environmental specialists confirm that areas near the uranium dumps have abnormally high levels of ambient radiation. Villagers admit that they visit the unguarded dump site to pick up scrap metal for resale. That poses dangers to the unwitting buyers who use the recycled steel for building houses and who may suffer health consequences years later. The Tajik authorities merely monitor radiation levels, as they do not have the funds to clean up areas like this.'

The foreign minister went on: 'In Taboshar, a former centre of uranium mining and milling, a hill of more than one million tonnes of processed residue tailings lies unprotected, vulnerable to erosion by wind and rain. Animals drink from pools of water that gather at the foot of the hill when seasonal rains fall, and children play around it. Some material from the tailings sites has also been used in home construction.

'Tajikistan is ill-equipped to undertake, on its own, the task of securing the tailings legacy. Vostokredmet, the company which mined uranium ore in the Soviet period, announced in October 2006 that the process of documenting radioactive waste pits had begun, and plans were being made to bury them safely. Our ecologists say plans to seek donor funding to bury radioactive waste in the north of the country are long overdue, and warn that local people are still disturbing contaminated earth in the area.'

Mr Zarifi spread his hands in a pleading gesture: 'Tajikistan would require hundreds of millions of dollars to decontaminate ten of the abandoned mines. The Organisation for Security and Cooperation in Europe plans to aid Tajikistan in working out a technical project to decontaminate these mines, and is calling on sponsors such as the International Atomic Energy Agency and NATO for funds.' The foreign minister said that he certainly hoped they would be successful.

But Tajikistan's problems do not simply relate to uranium tailings dumps. As an upstream nation with massive water resources, Stalin ordered the construction of gigantic reservoirs to hold water in the winter so that it could be released in the spring and summer to irrigate the cotton and rice plantations in the downstream republics like Uzbekistan, Kazakhstan and Turkmenistan. The damming of major rivers upstream, like the Amu Darya, and the squandering of scarce water resources in badly designed irrigation systems downstream, started the process that caused the virtual disappearance of an inland sea and led in turn to tensions between neighbours over trans-boundary water resources.

CHAPTER SEVEN

✯☭

The Aral Sea

The steel ribs of the old hull lie rusting on the sand. A ball of tumbleweed blows gently past in the stiff breeze. On either side, other rusty hulks form a solemn procession across the desert. In this eerie place, there is only the sound of the wind, which twists and furls the salty sand into mini whirlwinds, arching around the rotting wrecks of abandoned fishing boats. In the searing heat, the salt sticks to your skin and leaves a bitter taste on the lips. Only a profusion of shells and coral underfoot discloses the fact that this ships' graveyard was once a busy fishing port for up to 500 vessels. Today it is a tragic monument to man's greed and stupidity; a monument to Stalin's war against nature.

This was the sight that greeted me on my visit to Muynak in the spring of 2010. Together with my interpreter, Anna Dmitrijewa, and a senior official from the Ministry of Foreign Affairs in Tashkent, I embarked on a two-hour flight from the Uzbek capital to Nukus, the regional capital of the autonomous republic of Karakalpakstan.

The plane was a fairly modern turbo-prop, but the minute we hit a very slight bit of turbulence, the middle-aged woman in a colourful headscarf sitting on the other side of the aisle from me, followed by most of the other passengers, began to throw up, noisily! At that precise moment, the two stewardesses appeared with a trolley from which they started to serve tinfoil containers filled with a sort of grey, fatty stew. That just about set me off as well! I was greatly relieved when we finally landed in Nukus.

Nukus developed from a small settlement in 1932 into a

pleasant, modern Soviet city with broad avenues and big public buildings. However, the city's isolation made it host to the Red Army's Chemical Research Institute, a major research and testing centre for chemical warfare weapons. Now, with the desiccation of the Aral Sea and the attendant toxic dust storms that rage across this remote part of the autonomous region of Karakalpakstan, the city is only a shadow of its former self.

We were met by the *akim* and several government officials. Before we set off for Muynak, they were keen to show me the art museum for which this remote Uzbek city is rightly famous. Founded by the Russian painter, archaeologist and collector Igor Savitsky, the museum is noted for its collection of modern Russian and Uzbek art from 1918 to 1935. Stalin tried his best to eliminate all non-Soviet and avant-garde art during this period, and sent most of the artists to the Gulag. Both Savitsky himself and the collection at Nukus survived because of the city's remoteness.

Savitsky began collecting sculptures and canvasses of banned 'decadent bourgeois art' during the 1920s and '30s, when many artists were exploring new directions and experimenting with art movements prevailing at that time in the West. Stalin, whose preferred style depicted workers toiling happily in the fields, cracked down hard on such artists, condemning many to long prison sentences or years of hard labour. But Savitsky, who had made Nukus his home after visiting it on an archaeological expedition, persuaded these artists and their relatives to entrust him with their outlawed paintings and sculptures, often retrieving them from hiding places under floorboards or in attics. He rolled the paintings up and carried them back to his home in Nukus, gradually building up a vast collection numbering many thousands of works, covering every conceivable art form of the time from Fauvism and Expressionism to Futurism and Constructivism.

Eventually the authorities in Nukus, realising they were sitting on a collection of global importance, allocated a special museum for their display. The Karakalpakstan State Museum of Art is now a veritable treasure trove, filling in the missing pieces from the art history of the Soviet collections during the entire Stalinist period.

I was shown around this astonishing museum by its director Marinika M. Babanazarova, a remarkable woman who has fought off attempts by some of the Western world's leading art dealers to purchase key works from the Savitsky collection, often offering tempting amounts of cash. She showed me one particular work which poignantly summed up the whole legend of the collection. It is a large oil painting of a stylised bull by an artist called Vladimir Lysenko. The painting depicts a bull with enormous horns charging headlong out of the canvas towards the viewer. Rather disconcertingly, the bull's eyes are two completely black orbs which become the startling focal point of the painting.

Mrs Babanazarova explained to me that Vladimir Lysenko, who lived in Tashkent, was afraid to exhibit his painting of the bull for some years after he completed the work because he felt it might fall foul of the Soviet censors. He was right. When *The Bull* was eventually put on display in the first 'Republican Exhibition of Fine Art Workers of Uzbekistan' in Tashkent, the censors accused Lysenko of having painted a parody of the USSR, depicting the huge strength of Stalin's empire in the form of a prancing bull with the two totally black and unseeing eyes becoming a satirical comment on the blindness of the state. For this heinous offence, in 1935 Lysenko was arrested and sentenced to exile in the Gulag. He didn't reappear until 1951, when he returned to Tashkent a broken and disillusioned man. His works are amongst the most striking in the museum's collection.

Following our visit to the art museum, we set off by car on a three-hour journey to the once thriving seaport of Muynak on the Aral Sea. In 1960, the rotting hulks which now litter the desiccated seabed once landed 30,000 tonnes of fish a year in the bustling harbour. These boats were tethered to the harbour wall. But one morning the tide went out and didn't come back in. The water in Muynak harbour, which had been 20 metres deep, simply vanished, taking with it the livelihoods of thousands of fishermen, fish processors and port workers. Today, you have to travel more than 100 miles from Muynak to reach the sea. Unbelievably, this global catastrophe did not take centuries to materialise. It happened in the course of one generation.

The Aral Sea used to be one of the largest inland water reservoirs in the world, covering more than 40,000 square miles. Only Lake Superior in North America and Lake Victoria in Africa were bigger. This rich oasis on the Silk Road was an abundant source of food for generations of farmers, merchants, hunters and craftsmen, who came to trade and buy fish from the scores of local fishermen who plied their trade in the river deltas, lagoons and shallow straits. The Aral Sea was home to more than 38 species of fish, and its surrounding forests and hinterland teemed with a rich diversity of birds and wildlife, including deer, gazelle, Asiatic cheetah, lynx and even Caspian tigers.

All of that has gone. Now a salty desert stretches further than the eye can see. The searing summer temperatures and sharp winds have whipped up dust storms which can regularly deposit millions of tonnes of sand across hundreds of miles of neighbouring farmland. The disappearance of the Aral Sea has had a huge impact on climate change in the area. The rapidly extending desert has caused temperatures to rise in summer, while rainfall has decreased. In winter, severe frosts, which can see temperatures falling to $-40°C$, cause untold damage to agricultural crops.

How did this global ecological catastrophe happen? Sadly, it was entirely man-made. Those who think that mankind cannot cause climate change should visit the Aral Sea. It was here, in the rich, cotton-growing areas of Uzbekistan, that the Soviets decided in the 1950s to build a network of canals and irrigation channels to provide water for a massive extension of the cotton crop. They had visions of this area meeting all of the cotton requirements of the USSR.

The water for this new ambitious complex of irrigation channels was diverted from the two main rivers which serve the Aral Sea, the Amu Darya and the Syr Darya. The fields of cotton flourished, and when they faced a new threat from swarms of hungry locusts, Stalin ordered the crops to be sprayed with heavy doses of DDT and other poisonous pesticides. At harvest time, the cotton fields were sprayed with a desiccant, like Agent Orange, used by the Americans in Vietnam, to remove the green foliage from the cotton plants, so that the balls of cotton could be cleanly collected.

By the early 1970s, locals could already see the looming disaster. The two great rivers had been reduced to a mere trickle of water. The melting snows and rainfall from the mountains of Tajikistan and Kazakhstan were no longer reaching the sea. It was like pulling a plug out of a bath. Within only a few years, the mean water level had fallen from an average of over 53 metres to only 26 metres. The sea shrank to one-tenth of its former size, retreating quickly across the desert, leaving a wilderness of desolation in its wake and emptying hundreds of lakes and waterways along the course of the rivers that fed it.

Today, toxic dust storms course around the whole of Karakalpakstan. Agricultural land has been ruined. The food chain and local water supplies have been contaminated by salt and pesticides. Humans and farm animals are born with severe handicaps and disease. Illness is rife, and little tangible help seems to come from the international community. There is a standing joke in Nukus, the capital of Karakalpakstan, that if every consultant from the West who has visited the Aral Sea had brought a bucket of water, the sea would now have been filled up again. They are fed up with consultants and their endless reports in Nukus. They want action.

The Aral Sea is located between the Karakum and Kyzylkum deserts. In the Turkic languages, 'Aral' literally translates as 'island', and the sea is littered with countless small islands and shallow straits. The Aral Sea Basin drains approximately 1.8 million square kilometres and covers the entire territory of Uzbekistan and Tajikistan, the southern areas of Kazakhstan, the western and central parts of Kyrgyzstan, the eastern regions of Turkmenistan, part of northern Afghanistan and even north-eastern Iran.

Due to its size, the Aral Sea Basin possesses a range of diverse ecosystems, including the large Tien Shan and Pamir mountain ranges, the desert zones of the Karakum and Kyzylkum deserts, the fertile oases in the Syr Darya and Amu Darya rivers, and the Aral Sea water body itself.

The majority of freshwater entering the Aral Sea comes from glacial melt-waters in the Tien Shan and Pamir mountain ranges, and can be described as a terminal lake. In other words, it has a surface inflow but no surface outflow, meaning that the balance between inflows coming from the Amu Darya and Syr Darya rivers and net evaporation determine its level.

As the Aral is so expansive, the climate around it varies depending on the geographical location. Northern areas are temperate, with dry steppe and semi-desert landscapes, while the south is mostly desert, with a sub-tropical climate. Most of the 40 million population of the Aral Sea Basin tend to live near water sources.

The territory directly adjacent to the Aral Sea is called Priaraliye in Russian and has been an ecological disaster zone since 1989 – one of the worst the world has seen and a prime example of how unsustainable practices lead to environmental catastrophe. The severe environmental degradation became evident during the Soviet regime, but only came to international attention after its collapse.

In June 2010, I met Shavkat Hamraev, deputy minister of agriculture and water resources in Tashkent, Uzbekistan. Mr Hamraev told me that the UN Secretary General Ban Ki Moon was very shocked by what he saw in April, when he visited the Aral Sea in Uzbekistan, and promised to bring the Aral Sea to the attention of the international community.

In response to my criticism that Uzbekistan was still engaged in a cotton monoculture, Mr Hamraev said: 'Water utilisation by cotton crops in the area is down 30 per cent over the past 20 years, while the population has increased by 1.5 per cent. One hectare of land now uses 30 per cent less water. High evaporation is being caused by climate change in the area. As a result, cotton production is down. Farmers are growing more grain now, which doesn't require so much water. In 1985, yields of cotton per year were around 6 million tonnes. That has fallen to 3.5 million tonnes. Grain, meanwhile, has gone up from 1 million to 7 million tonnes per year, mostly wheat, 10 to 15 per cent of which is exported and the rest used in Uzbekistan. Yields have risen from 1 tonne per hectare to 4 or 5 tonnes per hectare.'

The deputy minister walked over to a map on his office wall. Pointing to an area between Nukus and Muynak, he said: 'Rice production in this area, which is a heavy water user, is down from 2 million to 1 million tonnes per year. We are studying less water-intensive crops and trying to encourage farmers in that direction. We are also reorganising canals and irrigation channels and looking at water droplet use. We are spending up to $100 million per year on improving soil quality in the region.'

I said that many people I had spoken to were critical of Uzbekistan's record on water conservation. Mr Hamraev replied: 'We have suffered much criticism for wasting water in Uzbekistan, but each year things are improving. We have factories now producing cement which is used to seal leaks in canals. Metal flumes and pipes have been introduced for saving water. Only 5 per cent of water is now lost. We have rich agricultural production in Uzbekistan and we are self-sufficient in wheat, poultry and meat. This is all due to good water management. Despite the financial crisis, our budget for irrigation has not been reduced.'

The minister was less sympathetic when it came to discussing hydroelectric power (HEP) projects in neighbouring countries. 'We do not agree with upstream HEP projects unless they have been properly evaluated by international assessors,' he snapped. 'A recent conference was held in Dushanbe on 9/10 June where the Koran was quoted on water use: "No one can have jurisdiction over water use . . . only God!" Uzbekistan has respected all international agreements. We are members of the World Water Council (WWC), the International Commission for Irrigation and the International Water Council. We would have no objection to ten small HEP projects in Kyrgyzstan, as they would not affect water use in the same way. The key priority is to satisfy, first and foremost, drinking water requirements. Only then can we look at irrigation, HEP schemes and large reservoirs.'

Despite the minister's assertions, Uzbekistan is still the world's third largest cotton exporter and earns around US $1 billion annually from the trade. Cotton production in Uzbekistan remains one of the most exploitative enterprises in the world and, since the time of Stalin, the government has routinely compelled hundreds of thousands of children to work as labourers in the country's annual harvest. Some analysts suggest between one and two million school-age children are forced to pick cotton every year.

Children as young as six years old, but mostly 11 and up, can be dispatched to the cotton fields for two months a year. Cotton picking is arduous labour, with each child given a daily quota of several dozen kilos that they must fulfil. Children receive little or no reimbursement for their labour.

Journalists and human-rights defenders exposing the issue have been subject to harassment and arrest, and independent monitoring is very difficult. Since 2007, international retail names including Tesco, Walmart-Asda and C&A have publicly condemned the use of forced child labour and rejected Uzbek cotton.

Cotton has such strategic significance for the Uzbek economy that Soviet-style production quotas are rigorously enforced. Despite some limited reforms, the government retains rigid control over the way in which cotton is grown, harvested and traded, and inputs such as chemical fertilisers and pesticides are state-controlled.

Prime Minister Shavkat Miriziyaev, who has control over the agricultural sector, reportedly convenes conference calls every 15 days in which he instructs local councils and farmers when to begin tasks such as seeding, weeding or applying pesticides.

In theory, farmers are private operators, but in reality they remain under the direct control of the state because they hold their land on a long lease rather than having outright ownership. They are also compelled to sell their crop to government-owned monopoly trading firms at prices far below the market rate. Selling cotton privately is treated as an illegal act. In this respect, nothing much has changed since Soviet times.

Since farmers have to bear all the production costs – which they pay for at free market prices – they cannot make much of a living from cotton, and even the larger leaseholders cannot afford to offer decent wages. The state even sends formal letters to farmers threatening them with court action if they fail to meet production targets. Unique to the Uzbek cotton industry is that farmers who grow almost the entire cotton crop are unable to profit from their work and remain indebted to the government, effectively making them bonded labourers.

By law, Uzbekistan's farmers must sell their cotton to the state-controlled company Uzkhlopkoprom that operates all of Uzbekistan's cotton gins. The price paid by the government is about one-third of the international market price, but farmers receive even less – as little as one-tenth of the global price – because their high-grade cotton is often judged as low grade. Uzbekistan owns 51 per cent of Uzkhlopkoprom. Information has never been publicly released about who owns the other 49 per cent.

Cotton holds a particular strategic importance to the state, and forced labour orchestrated by the government, and implemented by public employees and local authorities, is an integral part of production. Such is the strategic importance and public visibility of the cotton crop that in mid-October, during Uzbek President Karimov's trip to the Ferghana Valley, local authorities ordered the harvested fields to be 'decorated' with boxes of harvested cotton to create the illusion of fields overflowing with cotton.

In the Soviet era, mechanised cotton harvesters were used extensively, but a chronic failure by the state to invest in equipment means that there is not a single domestic harvester in use today. They were too costly to buy and maintain, and farmers soon realised that hand-picking was more cost-effective. A specialist from the International Cotton Advisory Council has previously estimated that the cotton harvest would actually require approximately 3,000 mechanical harvesters, which would involve a total investment of US $800 million.

The conditions in which many of Uzbekistan's cotton-pickers work can be characterised as forced labour. The work is arduous, and the climate is unforgiving. Physical abuse and threats keep children in the fields. The threat of expulsion is a major fear, and school grades suffer if the children do not meet their set quotas. But it is not only children who pick cotton. Teachers, police instructors, market traders, doctors, hospital staff and soldiers are all compelled to pick cotton during the harvest season.

In Nukus, the main centre for the growing of cotton and rice, I met Ubbiniyaz Ashirbekov, director of the Nukus branch of IFAS, the International Fund for Saving the Aral Sea. He told me: 'Water in the Aral Sea has retreated 150km from Muynak. We currently get around 4.5 billion cubic kilometres of water from the Amu Darya. If the Tajiks build their huge hydropower project on the Vakhsh River, known as Rogun, they say that it will take seven years to fill the reservoir with 22 billion cubic kilometres of water. This means the supply to the Aral Sea will be cut by this amount over those seven years. We cannot allow this to happen, or 200,000 to 250,000 hectares of land will be left without water. Our cotton crops will suffer. Wheat lands will be barren. Upstream countries are cultivating new agricultural land at our expense.'

He continued: 'The Tajiks could effectively cut off our water supply for two years with their reservoir projects. We are also alarmed at the Golden Century Lake in Turkmenistan. The water they are collecting has a very high salinity level. It will evaporate, and salt pans will be created. It could end up being a bigger catastrophe than the Aral Sea in 100 years' time.

'A study in 1989–91 concluded that 75 million to 120 million tonnes of toxic dust is being deposited on our land every year. We need urgent help to resolve the issue of salt-spread and toxic dust storms. The impact on agriculture, biodiversity, and meat and grain production is horrendous. The solution is part reconstruction of the ecosystem, planting salt-resistant plants and trees. But there are vast areas to cover, and in some of these areas nothing will ever be able to grow again. However, we could plant 350,000 hectares of trees around the Aral Sea Basin, but this will cost approximately US $1,000 per hectare. The economy of our country is currently in transition, and there are no funds available. But we only have another 50 years in which to arrest the desiccation. After that, there will be no way back.'

The Amu Darya and Syr Darya Rivers

In 1558, an English explorer called Anthony Jenkinson writing about this region said: 'The water that serveth all that country is drawn by ditches out of the River Oxus, into the great destruction of the said river, for which it cause it falleth not into the Caspian Sea as it hath done in times past, and in short time all that land is like to be destroyed, and to become a wilderness for want of water, when the river of Oxus shall fail.'

It took another four centuries and the advent of the Soviet empire to make Jenkinson's prediction come true.

The Amu Darya and Syr Darya are the two largest rivers in Central Asia, and each of the five independent republics has a stake in at least one of the rivers. Both rivers feed the Aral Sea, and together they make up the Aral Sea Basin.

The Amu Darya River

The Amu Darya River, in ancient times known as the 'Oxus', originates from glaciers and snowmelt in the Pamir Mountains, and is formed at the confluence of the Panj and Vakhsh at the Tajik–Afghan border. The Amu Darya is approximately 2,400–2,600km long and forms part of the borders between Afghanistan and Tajikistan, Afghanistan and Uzbekistan, Afghanistan and Turkmenistan, and Uzbekistan and Turkmenistan. Additionally, the Amu Darya Basin encroaches on both Iran and Kyrgyzstan. In its final reaches, the Amu Darya flows through the Karakum Desert before crossing Karakalpakstan and entering the southern or 'Large' Aral Sea.

Tajikistan contributes 80 per cent of the flow generated in the Amu Darya followed by Afghanistan with 8 per cent and Uzbekistan 6 per cent. Kyrgyzstan, Turkmenistan and Iran each contribute under 3 per cent. Despite contributing relatively little of the water resources in the basin, Turkmenistan and Uzbekistan are the major water users, mainly for irrigated agriculture. In Turkmenistan, the Karakum Canal (1,400km long) carries water from the Amu Darya near Kelif across Turkmenistan to Ashkhabad, and supplements the flow of the Tejen and Murgab rivers. Quite apart from the effects of the Aral Sea disaster, one of the main causes of tension between water users in the basin is the overuse of water for irrigation.

The greatest tensions currently exist between Turkmenistan and Uzbekistan, and there have even been reports of military skirmishes. Uzbekistan is concerned that Turkmenistan's overuse of water will create shortages which will impact on agricultural production and even on the availability of drinking water in Uzbekistan.

The Syr Darya River

The Syr Darya, formerly known as 'Jaxartes' or 'Yaxarte', is even more heavily regulated than the Amu Darya, with many hydroelectric facilities, reservoirs and canals lining its tributaries. Although it is the longest river in Central Asia (approx. 3,000km, including the Naryn River), the Syr Darya carries less water than the Amu Darya, and courses through Kyrgyzstan, Tajikistan, Uzbekistan and Kazakhstan. The Syr Darya flows from the Tien Shan Mountains, and on the years that it reaches the Aral Sea, it feeds the Northern or 'Little Aral'. Kyrgyzstan contributes 74 per cent of its surface flow, followed by Kazakhstan (12 per cent), Uzbekistan (11 per cent) and Tajikistan (3 per cent).

Both the Amu Darya and Syr Darya river basins have an extensive network of dams, reservoirs and irrigation canals which form one of the most complex basins in the world. Largely due to poor water management practices and the desiccation of the Aral Sea, the watersheds of the Amu Darya and Syr Darya are now distinct and separate, at least as far as surface waters are concerned. Their surface waters no longer intermingle at the end of their courses and the Aral Sea is therefore no longer a contiguous body of water.

Aside from the densely populated areas of northern Kazakhstan, most of Central Asia's 60 million people reside in the area drained by the Amu and Syr Darya rivers. The mountainous upstream states of Kyrgyzstan and Tajikistan are largely dependent upon agricultural economies, whereas the downstream states of Kazakhstan, Turkmenistan and Uzbekistan show a greater balance between agriculture and industry.

In terms of national security, all Central Asian states depend on the Syr Darya and Amu Darya. Tajikistan, Kyrgyzstan and Uzbekistan are the major actors in strategic terms, and considerably affect the way in which water resources are managed in the region. Turkmenistan and Kazakhstan are very much dependent on the rivers for agricultural development but are not in strong strategic positions.

CHAPTER EIGHT

Desiccation

In 1960, the Aral Sea was the world's fourth largest inland water body, covering more than 40,000 square miles. Although it was a brackish-water body (salinity one-third less than the ocean), the Aral was inhabited mainly by freshwater species of fish. It was home to a large fishing industry as well as functioning as a key regional transportation route. The vast deltas of the Syr Darya and Amu Darya rivers sustained a diverse range of flora and fauna, and supported irrigated agriculture, animal husbandry, hunting, trapping, fishing and livestock farming.

Although the Aral has naturally expanded and shrunk throughout history, since the 1960s reduced inflow has caused unrelenting desiccation. Rather than severe drought or natural climatic change, the reduced inflow was a direct result of irrational water use and overexploitation of water resources from the two main feeder rivers, the Amu Darya and the Syr Darya. The Soviet leadership's imposition of 'modern' cotton farming methods and the associated diversion of waters away from the Aral Sea for irrigating water-intensive crops such as cotton were the principal causes of this environmental catastrophe.

One of the key diversions of water away from the sea occurred with the construction of the Karakum Canal, which is still the world's biggest irrigation canal. Started in 1954 and completed in 1988, the canal in Turkmenistan cuts through the Karakum Desert and stretches for 1,400km, all the way to the Caspian Sea. To fill this canal and irrigate the desert areas, all the water comes from the Amu Darya River, which is the primary source of water for the Large Aral Sea. To exacerbate the problem, this huge canal was poorly designed and loses around 50 per cent of its water along its length. However, it helped to irrigate great new tracts of cotton and was quickly followed by scores of other new irrigation channels, all of which diverted water from the Amu Darya and the Syr Darya rivers. At the time of its construction, the Karakum Canal was hailed as a huge success because of its 'bold transformation of nature'. It was initially named the V.I. Lenin Canal, and the leading experts who designed it were awarded the Lenin Prize.

The disastrous consequences of such poor water-management techniques became fully evident in the 1980s when the Aral Sea separated into two distinct water bodies – the northern 'Small Aral', which is fed by the Syr Darya, and the southern 'Large Aral', which is fed by the Amu Darya River. By 1987, the discharge was down to a trickle and on three separate occasions between 1981 and 1988 there was zero input from these rivers. By 1989, the shoreline was reduced by 480km, and in some places the coast had receded by over 100km, leaving many former coastal towns and fishing villages surrounded by the Aralkum Desert.

Even though there were predictions that large-scale irrigation might lead to a shrinking sea, the Soviets were not worried, as they believed that a smaller Aral Sea would provide more fertile land to grow their 'white gold' – cotton. It is important to note that the Soviets were not the first to have grandiose plans for the Aral Sea Basin. Historic records show that even back in the reign of Tsar Peter the Great, engineers were toying with the idea of draining the Aral Sea to create what they thought would be a vast fertile wetland, capable of supplying Mother Russia with endless agricultural produce.

Prior to 1997, the Large and Small Aral were connected by a channel which flows from the small to the large. However the flow was erratic at best, and in 1997 Kazakhstan built a permanent dike to separate the two lakes.

As of 2008, the surface of the sea had shrunk to less than half of its 1960 size; its level has fallen by 23 metres and the salinity of the water had increased significantly. The area of the dried-up sea bed now exceeds 4 million hectares.

The desiccation was so severe because the Soviet regime increased the area of irrigated land with little consideration for the local environment. The shrunken sea is the starkest symbol of the USSR's poor water resource management, and highlights that the Soviets viewed nature as something to be marshalled and directed by elaborate engineering. Thus, Moscow spent billions on building dams and canals across Central Asia to increase the area of irrigated land without concern about environmental damage.

In Soviet times, there was also an abundance of forced labour which could be deployed to hand-dig irrigation channels and even huge canals. The infamous Belomor Canal gave the Soviet Union a shipping canal from the White Sea to the Baltic, at a cost of at least 11,000 lives and the use of the forced labour of over 100,000 political prisoners. Even though the canal is narrow, shallow and seasonal, it is still in use, especially for the transport of oil.

Although the Soviets claimed such labour camps were a means of 'rehabilitation', this was, essentially, slavery. Belomor is one of the more notorious canals built by forced labour in the USSR, but the vast Karakum Canal in Turkmenistan and many of the smaller canals and irrigation channels in Uzbekistan were built in this way. The infamous 270km-long Ferghana–Stalin Canal was dug by hand by 180,000 so-called 'volunteers' in only 45 days. This energetic construction programme accelerated the amount of water being drained from the main Amu Darya and Syr Darya river systems and led to the gradual desiccation of the Aral Sea.

It took so long to raise awareness of the desiccation because the environmental issues in the region can be classified as a 'creeping environmental problem' where incremental changes accumulate over time at a slow rate. Thus degradation generally is imperceptible and not recognised until a crisis has emerged, meaning that behavioural changes are not initiated and environmental problems are neglected.

By continuously extending the irrigation systems into the

deserts, the Soviets overwhelmed the natural balance of the basin and kept increasing the amount of water being diverted from the Amu Darya and Syr Darya rivers. Due to the arid climate and desert environment that was being irrigated, much of this water was being lost through evaporation rather than being returned to the rivers as run-off.

The desiccation was exacerbated because the irrigation systems were unlined and uncovered, and therefore extremely inefficient and poorly maintained. Despite the large diversions of water from the Amu Darya and Syr Darya rivers, it is estimated that 50 to 90 per cent of water diverted for irrigation never reaches the crops due to poorly designed irrigation canals and the fact that water users have historically received water for free. The consequences of the desiccation are far-reaching and have impacted on fish, living organisms, soil quality, air quality, ground and surface water quality, environmental sustainability and, of course, the human population of the basin.

Consequences of the Aral Sea Desiccation

Impacts on Wildlife

The ecosystems of the Amu Darya and Syr Darya deltas have been severely affected by the Aral Sea disaster: they have become major casualties of Stalin's war against nature. Both deltas used to encompass rich vegetation which served as precious wildlife habitats, but the continual desiccation of the sea caused a contraction of the flooded area and shortened the duration of flooding in parts of the deltas. The shrinking sea also contributed to severe climate change in a band up to 100km wide along the former shoreline in Kazakhstan and Uzbekistan. Maritime conditions have now been replaced by more continental and desert-type conditions, meaning summers have warmed and winters cooled, spring frosts are later and autumn frosts earlier, humidity is lower and the growing season shorter.

Prior to the construction of the large irrigation channels, over 70 species of mammals and 319 species of birds lived in the river deltas. Today, only around 32 species of mammals and 160 of birds are left. All 24 native species of Aral fish have now completely disappeared from the sea. Zones that provide prime habitat for a variety of permanent and migratory water-fowl, a number of which are already endangered, have been destroyed by the environmental degradation caused by toxic dust storms and salt. Reduced river flows and declining groundwater levels that affected wildlife and vegetation have also caused serious salinisation, which is a key contributor of desertification.

Impacts on Land

Salinisation in the Aral Sea Basin has occurred both in water and on land as a result of the poor water management techniques used in the region. Water salinity has increased dramatically since large-scale irrigation practices began, and now any water that does reach the sea has elevated salinity because repeated usage of water in the middle and upper courses of the rivers has caused widespread leaching of salts from the soil. Excessive irrigation in the basin has caused the mobilisation of deep salt reserves in the soil, raised the water table and waterlogged fields. The soil is now so saline that the once lucrative cotton industry is struggling.

As the Aral Sea has shrunk, the salts suspended in the water have been deposited and accumulated on the former seabed. Salt accumulation has resulted in salt pans, where nothing will grow and indigenous species such as tugay vegetation have been devastated. When salts in the form of dust settle on vegetation and crops, they cause a reduction in growth and can also harm animals that rely on the plants as a food source.

The effects of salinisation have devastated vast areas of the basin. Ninety-five per cent of irrigated lands in Turkmenistan and 16 per cent in Tajikistan suffer salinisation, and in Kazakhstan approximately 30 per cent of agricultural lands are salinated, waterlogged or at risk. Salinisation and desertification in Central Asia are so severe because not only did the Soviet regime over-irrigate the land, but it also overused pesticides and chemicals to increase productivity and economic yield. Years of excessive pesticide and herbicide use mean that the salt pans are not only highly saline but also heavily contaminated, and now the soil and groundwater in the Aral Sea Basin is extremely toxic to both humans and wildlife.

Fields in the region were intensively sprayed with persistent pesticides, often from aeroplanes that flew over villages, spraying villagers and cotton field workers. To increase cotton yields, large quantities of persistent organic pollutants (POPs) were used, including DDTs, HCH, lindane and toxaphene. While DDT was banned in 1972, there is evidence that its use persisted in the Aral Sea Basin. Toxophene was banned in the West due to its carcinogenic and mutagenic effects, and it has largely been replaced by lindane in Central Asia. As excessive irrigation practices continued, salinity of soils and water increased. To improve crop yields, more water, fertiliser and pesticides were applied, which is why the salt pans on the former Aral seabed are so contaminated with the residue from pesticides and POPs.

The accumulated salts and pesticides are easily transported by the strong prevailing winds in the basin. The exposed sediments are picked up and carried by frequent dust storms and have been disseminated as a white powder all over the region, exposing the whole ecosystem to toxic components. Since the mid-1970s, satellite images have shown major dust plumes extending as far as 500km downwind, dropping dust and salt over a considerable area adjacent to the sea in Uzbekistan, Kazakhstan and even Turkmenistan, which does not even border the sea.

These large dust storms occur on average at least ten times annually and deposit tens of millions of tonnes of contaminated dust on the surrounding land every year. It is estimated that the increasingly frequent storms displace approximately 43 million tonnes of dust per year and, due to the northerly prevailing winds in the region, Karakalpakstan and Turkmenistan are the worst-affected areas. The dust also has serious consequences for the health of the entire population living in the Aral Sea Basin.

Impacts on Human Health

The deteriorating health situation in many of the Basin countries is intricately linked with the worsening ecological and economical situations of the region. Diseases have increased, particularly rates of anaemia, tuberculosis, kidney and liver diseases, respiratory infections, allergies and cancers, and their frequency far exceeds the former USSR and present-day Russia. Poor agricultural practices have led to contaminated drinking water, which has high levels of salt, pesticides, heavy metals and bacteria. Consequently, infant mortality rates are high, life expectancy has decreased by 20 years, and cancer, gastrointestinal problems, typhoid and hepatitis are widespread.

Although environmental degradation has health consequences for all residents in the basin, it is undoubtedly the 1.5 million population of the Karakalpakstan Republic of Uzbekistan that is most affected. Ninety-six per cent of Karakalpaks live in the polluted area surrounding the Aral Sea.

The people of Karakalpakstan, who are culturally and ethnically distinct from the rest of Uzbekistan, have borne much of the brunt of the ecological disaster. Today, settlements such as Muynak, which used to sit on the shores of the Aral, are over 100km from the water, and are some of the most chronically sick places on earth. In Karakalpakstan, rates of anaemia, cancer and tuberculosis far exceed those of the rest of Central Asia, and the infant mortality rate is more than 100 per 1,000. In other areas of the Aral Sea Basin, the infant mortality rate is estimated to have increased from about 25 per 1,000 in 1950 to 70–100 per 1,000 in 1996. Low birth weight, growth retardation and psychoneural retardation are also much more prevalent than normal.

Besides Karakalpakstan, areas of Kazakhstan which border the Aral Sea also experience high infant mortality, sharp increases in oesophageal cancers and typhoid, outbreaks of viral hepatitis, gastrointestinal problems, and high rates of congenital deformation. The Kzyl-Orda Oblast in Kazakhstan and the Dashauz Oblast of Turkmenistan have suffered more than most areas. In the former, average life expectancy has declined from 64 to 51 years.

Women are among the most vulnerable to health problems associated with environmental degradation, and significantly high levels of organic pesticides like HCB, HCH, DDE and DDT have been found in the plasma of pregnant women, again far higher than in EU countries. Child health in the Aral region is also poor, with high mortality, a high rate of chronic disease and retarded mental and physical development.

Bacterial contamination of drinking water is pervasive in the Basin states because desiccation of the sea caused the groundwater table to rise, which then became contaminated with high levels of salts and other minerals. Contamination of potable water means that there is a serious shortage of drinking water, and its poor quality has led to very high rates of typhoid, paratyphoid, viral hepatitis and dysentery. Tuberculosis, anaemia, and liver and kidney ailments are widespread, and it is assumed that the latter is related to the high salt content of drinking water, amongst other things.

A local doctor I sat next to on the flight back to Tashkent from Nukus told me that the prevailing poor health conditions are partly due to an inadequate health care infrastructure since the collapse of the Soviet Union, and partly due to socio and ecological factors. She said that inadequate nutrition, poor sanitation and the virtual collapse of the health care system after independence have all contributed to the local health crisis, as well, of course, as pollution from Soviet agriculture and associated industries.

Impacts on the Fishing Industry

Environmental degradation is closely linked to economic losses in the Aral Sea Basin and has resulted in the loss of thousands of jobs, particularly in fishing-related activities. The once-booming Aral fishing industries developed by Kazakhstan and Uzbekistan have disappeared, owing to rising salinity and loss of shallow spawning and feeding areas. Although some native species of fish still survive in the Syr Darya and Amu Darya, the Aral salmon is extinct.

In the decade after independence, nearly every regional

Left. The first Soviet nuclear explosion, nicknamed 'Joe-1' by the Americans. (Source: Atomic Archive 2010)

Below. One of the many radioactive craters that have ruined farmland and water bodies in the Polygon. (Photo: Kimberley Joseph)

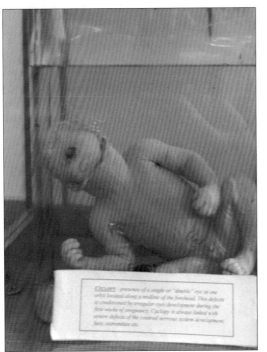

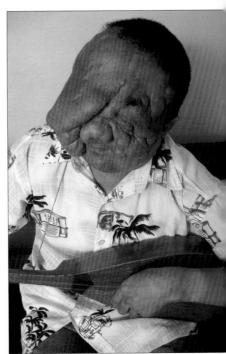

A boy born to a Soviet pilot and his wife who were working on the nuclear tests in the Polygon, with a single eye in the centre of his forehead – a perfect Cyclops. This severely deformed baby suffered as a result of the radiation his parents had been exposed to. He survived for 24 hours after birth. (Photo: Struan Stevenson)

Some of the citizens of the Polygon have suffered terrible deformities due to the legacy of nuclear testing. Berik Sadykov from Znamenka village. (Photo: Kimberley Joseph)

Angry Kazakh villagers from Sarzhal in the Polygon have called for more help for victims of the Soviet nuclear tests. (Photo: Kimberley Joseph)

The Tsar Bomba – the largest, most powerful nuclear weapon ever detonated.

ruan Stevenson MEP, and UN Secretary General Ban Ki-Moon during a visit to Kurchatov.
Photo: Kamila Magzieva)

Struan Stevenson Street in Znamenka. (Photo: Struan Stevenson)

Struan Stevenson with parents and children at the opening of the new Urdzhar School for Handicapped Children. (Photo: Struan Stevenson)

The rapidly receding shoreline of the Aral Sea has left many fishing boats strewn across the former seabed. (Photo: Struan Stevenson)

Struan on one of the rotting hulks at Muynak harbour, which is now over 100km from the sea. (Photo: Struan Stevenson)

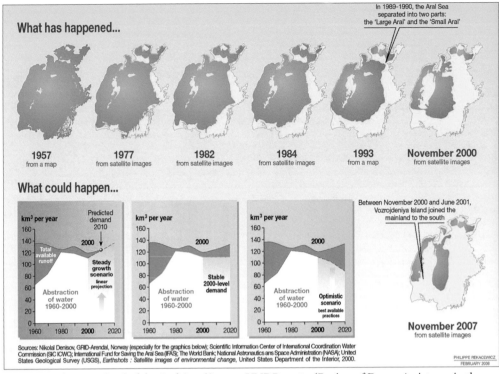

What has happened...

In 1989-1990, the Aral Sea separated into two parts: the 'Large Aral' and the 'Small Aral'

| 1957 from a map | 1977 from satellite images | 1982 from satellite images | 1984 from satellite images | 1993 from a map | November 2000 from satellite images |

What could happen...

Between November 2000 and June 2001, Vozrojdeniya Island joined the mainland to the south

km³ per year
160
140 Total 2000 Predicted demand 2010
120 available runoff
100 Steady growth scenario
80
60 Abstraction linear projection
40 of water 1960-2000
20
0
1960 1980 2000 2020

km³ per year
160
140
120 2000
100
80 Stable 2000-level demand
60 Abstraction of water 1960-2000
40
20
0
1960 1980 2000 2020

km³ per year
160
140
120 2000
100
80
60 Abstraction of water 1960-2000 Optimistic scenario best available practices
40
20
0
1960 1980 2000 2020

November 2007 from satellite images

Sources: Nikolaï Denisov, GRID-Arendal, Norway (especially for the graphics below); Scientific Information Center of International Coordination Water Commission (SIC ICWC); International Fund for Saving the Aral Sea (IFAS); The World Bank; National Astronautics ans Space Administration (NASA); United States Geological Survey (USGS), Earthshots : Satellite images of environmental change, United States Department of the Interior, 2000.

PHILIPPE REKACEWICZ FEBRUARY 2008

Changes in the water level of the Aral Sea. (Source: UNDP 2008 – 'Review of Donor Assistance in the Aral Sea region 1995–2005')

With the wife of the akim of Muynak after being welcomed into the furnace-like interior of their home and naming their grandson 'William Wallace'. (Photo: Struan Stevenson)

The Nurek dam in Tajikistan, currently the tallest in the world. (Photo: Struan Stevenson)

Struan at the Nurek dam in Tajikistan, with the slogan 'Water Is Life' painted over a tunnel entrance. (Photo: Struan Stevenson)

Right. Struan with Tajik president Emomali Rahmon at the Rogun dam. When fully operational, the Rogun dam will supply electricity to Afghanistan and Pakistan, as well as Tajikistan. (Photo: Struan Stevenson)

Below. Struan with one of the less aggressive fighting dogs near Ashkhabad in Turkmenistan. (Photo: Struan Stevenson)

fishery went out of business because of the extinction of numerous species and the disappearance of the Aral Sea as a living body of water. In 1960, the commercial fishing catches were approximately 43,430 tonnes compared to 17,460 tonnes in 1970. In 1980, commercial fishing catches were non-existent. I was told that 40 to 60,000 fishermen have lost their livelihoods, devastating those areas of Karakalpakstan that used to rely on fishing for up to 50 per cent of their income. The loss of fisheries has also led to increased pollution, as any industries related to fishing have vacated the area, leaving their residues behind. Both the loss of jobs and the poor health situation has led to the creation of environmental refugees. It is estimated that more than 100,000 people were forced to leave their homes in the Aral Sea region after the sea died.

What Can Be Done?

First of all, as the local doctor assured me, Nukus needs a new hospital. She said the local population suffer the same symptoms as sufferers of HIV/Aids. Their immune systems have broken down, and they suffer liver and kidney failure and a variety of cancers affecting the throat, thyroid, stomach, liver and pancreas.

Secondly, pioneering work in Muynak needs to be encouraged and properly resourced. A new lake has been constructed and fish have been reintroduced. Large areas of land are being planted with salt-resistant varieties of tree. Thirdly, hydro-power projects and reservoir schemes in some of Uzbekistan's upstream neighbours in Central Asia must be re-evaluated to ensure they do not further reduce the water flow to the Aral Sea.

The West can help. The people of Karakalpakstan are the victims of the Cold War. They suffered at the hands of Stalin's dictatorial race for world domination. We can learn a lot from their plight. The Aral Sea is a tragic laboratory of climate change. The environment, ecology, biodiversity and life itself can be wiped out in a single generation in the pursuit of avarice. Let us learn the lessons of the Aral Sea, try to restore it to its former glory and vow never to allow such devastation to happen again, anywhere on our planet.

Let me end the sad tale of the Aral Sea on a lighter note. When I arrived in Muynak, the former fishing port now stranded in the middle of a windswept desert, I was met by the *akim*, who took me to see the rotting hulks of fishing vessels lying on the sandy wasteland. Afterwards he insisted on taking me to his home to meet his family. It was 40°C, and my heart sank when I was ushered into the *akim*'s home, which had a corrugated tin roof and windowless walls, heavily lined with woollen rugs and carpets. The temperature inside was like a furnace and, of course, there was no hint of air conditioning. Even the Uzbeks were fanning themselves furiously to try to keep cool.

I was accompanied by Anna Dmitrijewa – my assistant and interpreter – and two officials from the Uzbek Ministry of Foreign Affairs. The *akim* and his family had prepared a lavish feast in our honour. The centrepiece was the usual sheep's head from which the eyes were plucked and offered to me as a special delicacy. To wash them down, the *akim* poured copious glasses of vodka and proposed a series of toasts to his visitors from Europe. Suddenly, he announced that his daughter had just that morning given birth to a son. We all applauded. But his next remark almost made me choke on my vodka. My interpreter explained that he was asking me to name his grandson.

All eyes in the room were now turned in my direction. I had been sweating before, but now it was as if someone had turned up the thermostat. My mind was racing. What would be an appropriate name for a boy in a remote corner of Uzbekistan? I turned to my interpreter and asked her if she knew what the Uzbek was for 'Braveheart'. But before she could answer the *akim*'s eyes lit up and he shouted excitedly in English 'William Wallace'. He'd seen the movie.

The *akim* summoned his wife from the kitchen and gleefully announced that her new grandson was to be named William Wallace Balakbaev. I had to pose for photographs sandwiched between the *akim* and his wife. They depict me, my shirt stuck to my chest with sweat, while the *akim* exposes a huge array of gold-plated teeth in a wide grin. The *akim*'s wife, poor soul, appears rather crestfallen, no doubt trying to figure out how she will one day explain to her grandson how he came to be called William Wallace.

Refill the Aral Sea?

One potential way to save the Aral Sea is through more efficient irrigation systems. The Karakum canal, which is the centrepiece of the Soviet-built system of huge diversion canals, lacks concrete lining for much of its length. In all there are over 60 diversion canals that tap into the Amu Darya and Syr Darya rivers, most of which lack lining. Improving the canal system and overall irrigation efficiency may return as much as 20 cubic kilometres of water per year back to the Aral; not nearly enough, but far more than at present. However, the United Nations Development Programme suggested that inefficient irrigation and drainage infrastructure inherited from the Soviets requires approximately $23 billion of investment for rehabilitation.

Other ideas have included rekindling Soviet plans to divert water from Siberian rivers as a potential way of refilling the Aral Sea. When these diversionary plans were conceived by the Soviets, they did not anticipate the effect on the Caspian Sea level, which has steadily risen in recent years, inundating oil exploration sites and causing widespread pollution.

Some experts claim that draining up to 10 per cent of the water from the Volga River would not only help to alleviate the devastation of the Aral Sea, but also reduce the inflow into the Caspian Sea, maybe even slightly reducing the sea level and preventing the inundation of oil facilities and coastal communities.

Russian Prime Minister Vladimir Putin, Moscow Mayor Yuri Luzhkov and, predictably, Uzbek President Islam Karimov, have endorsed reviving the project. But building this canal, which would be up to 800km long, 200m wide and 16m deep, would cost billions and would tap into the Volga River in the Ural uplands, running along the contours of the earth to drain into the Aral basin and into the shrunken sea. Due of a lack of high-altitude barriers between the Volga watershed and the Aral basin, this canal would require few or even no pumping stations.

It is commonplace for environmentalists to ridicule this diversionary plan, as they feel that it would have knock-on impacts on the environment in Siberia, create environmental concerns in Siberia, cause further salinisation and disrupt many local ecosystems. However, research needs to ascertain whether the long-term benefit will outweigh the enormous costs. Is it worth it to reverse the biggest environmental disaster in history?

Meanwhile, the disappearing sea has had drastic consequences for another of Stalin's ill-considered environmental adventures.

CHAPTER NINE

☭

Vozrozhdeniye Island

One of the most serious consequences of Stalin's war against nature and the desiccation of the Aral Sea is the enlargement of Vozrozhdeniye Island. Vozrozhdeniye Island (also known as Resurrection or Rebirth Island) is located in the Large, or Southern, Aral Sea. The island is characterised by large, sparsely populated deserts with scrubby vegetation due to the arid climate and sandy soils. The northern third of the island, known as Merensay, is now Kazakh territory, but Kazakhstan has not yet put the land to economic use, as specialists remain concerned about environmental contamination. The rest of the island is controlled by the autonomous Karakalpakstan region of Uzbekistan.

In 1936, control over Vozrozhdeniye Island was transferred to the Soviet Defence Ministry for use by the Red Army's Scientific Medical Institute. That summer, an expedition of 100 people led by Professor Ivan Velikanov arrived on the island in the utmost secrecy. This team of researchers and scientists were provided with special ships and two aircraft, and reportedly conducted experiments involving the spread of tularaemia (a disease of rodents that can be spread to humans by insect bites) and related microorganisms. In late 1937, however, with Stalin suffering from one of his regular bouts of acute paranoia about spies and infiltrators, the expedition was evacuated from the island due to security fears. The KGB even arrested Professor Velikanov and some other specialists on suspicion of espionage.

The island remained closed for all intents and purposes during the Second World War, but in 1952, the Soviet government again decided to resume biological weapons testing there. A test site, officially referred to as 'Aralsk-7', was built in 1954 on Vozrozhdeniye. The Defence Ministry's Field Scientific Research Laboratory (PNIL) was stationed on Vozrozhdeniye Island and instructed to conduct the experiments. A military guard unit comprising several hundred soldiers was also based on the island, and reported to a larger unit based in Aralsk. The PNIL developed methods of biological defence and decontamination for Soviet troops. Specific items of military equipment, weapons and protective clothing were taken to the test site for exposure to biological weapons and decontamination before being mass produced and distributed to Soviet forces. Following the Soviet invasion of Afghanistan, military protective gear developed for Afghan conditions was tested at the PNIL on Vozrozhdeniye Island.

The top-secret site itself was divided into a testing complex in the southern part of the island and, upwind, a military settlement in the northern part, where officers – some with families – and soldiers lived. The settlement had barracks, residential houses, an elementary school, a nursery school, a cafeteria, warehouses and a power station. Staff and their families were subjected to regular immunisation and received special hardship payments and privileges. PNIL laboratory buildings, located near the residential area, possessed up-to-date equipment and a special biosafety shelter where personnel could hide in an emergency. Also located in the northern part of the island was Barkhan airport, which provided regular plane and helicopter connections to the mainland. There was also a harbour at Udobnaya Bay, where special fast patrol boats protected the island from intruders.

The open-air test site in the southern part of the island was used for studying the dissemination patterns of biological weapons agent aerosols, as well as methods for their detection and the effective range of aerosol bomblets with biological agents of different types. The testing grounds were equipped with an array of telephone poles with detectors mounted on them, spaced at intervals of one kilometre. Biological weapons agents tested at the Vozrozhdeniye site had been developed at

the Defence Ministry facilities in Kirov, Sverdlovsk and Za-
gorsk, and the Biopreparat centre in Stepnogorsk, and included
anthrax, tularaemia, brucellosis, plague, typhus, Q-fever, smallpox, botulinum toxin and Venezuelan equine encephalitis.

The experiments were conducted on horses, monkeys, sheep
and donkeys, and on laboratory animals such as white mice,
guinea pigs and hamsters. In addition to common pathogenic
strains, special strains developed for military purposes were
tested at the island. Bacterial simulants were also used to study
the dissemination of aerosol particles in the atmosphere. Researchers on Vozrozhdeniye Island used to joke that the condemned monkeys were the luckiest inhabitants of the Soviet
Union because they lived on fresh fruit like bananas, oranges
and apples, which were rare delicacies for most human residents of the Soviet Union at that time. But these laboratory
animals had to remain in prime condition right up to their last
breaths, usually taken strapped to a pole out in the killing fields
near the laboratories. Meanwhile, the cream of Soviet scientists
who were conducting the atmospheric tests had to live on
hunks of bread and fatty sausage.

The fact that the island's prevailing winds always blew
toward the south, away from the northern settlement, was
probably an important factor in designing the site. The biological weapons aerosol tests were also conducted in such a way
as to avoid contaminating the northern military settlement, and
a special service on the island was responsible for imposing
strict environmental controls. Nevertheless, the activities on
the secret island caused serious concerns among local residents
on the mainland because of repeated epidemics and the mass
deaths of animals and fish in the area. There were also sporadic
reports of individual cases of infectious disease in people who
spent time on the island.

Towards the end of the Soviet era in 1988, hundreds of
tonnes of anthrax were transported from other Soviet test sites
such as Stepnogorsk and dumped on Vozrozhdeniye. The
anthrax strains were hastily buried in drums or simply in
sandpits to which bleach was added. The Moscow authorities
did not allow Kazakhstan's public representatives to visit

Vozrozhdeniye Island until a first investigative commission went there in 1990.

A second Kazakhstan government commission, headed by the Kazakh minister of ecology and bio-resources, visited the island in 1992. In August that year, an independent expert commission of the Aral-Asia-Kazakhstan non-governmental organisation also visited. Meanwhile, the Russian military authorities were claiming that no offensive testing or research had ever been conducted on the island and that the site had only been used for the testing of defences against biological weapons.

The evacuation of Russian military personnel from Vozrozhdeniye Island began in 1991. All the PNIL specialists left, and the laboratories were mothballed. In January 1992, the Supreme Soviet of the newly independent Republic of Kazakhstan issued an edict 'On Urgent Measures for Radically Improving the Living Conditions of Aral Area Residents', which officially closed the Vozrozhdeniye military site.

In April 1992, Russian president Boris Yeltsin's edict no. 390, 'On Ensuring the Implementation of International Obligations Regarding Biological Weapons', ordered that all offensive bio-weapons programmes be shut down. Following this decree, the Russian government declared that the Vozrozhdeniye site was to be closed, the special structures would be dismantled, and within two to three years the island would be decontaminated and transferred to Kazakhstan's control.

In August 1995, specialists from the US Department of Defense visited Vozrozhdeniye Island and confirmed that the experimental field lab had been dismantled, the site's infrastructure destroyed and the military settlement abandoned. Members of the US team also reported that on arrival at the site of the crumbling former laboratories they were met with a chaotic sight. Test tubes and Petri dishes were scattered around the floors of the buildings next to rotting cages. Pictures still hung on some walls, and in one room they even found a bed with blankets and sheets still in place. It was as if the scientists had evacuated the place in a hurry. Some reports suggest that not much has changed to this day.

Despite warnings of the dangers, after the Russian authorities left Vozrozhdeniye Island in 1992, and in the wake of the economic crisis caused by the collapse of the Soviet Union, local residents of Kazakhstan and Uzbekistan flocked to the island to seize abandoned military equipment that the Russian forces had left behind. It is to be hoped that the looting occurred in the safer residential part of the island, as specialists remain concerned about environmental contamination.

The work conducted on Vozrozhdeniye was so secretive that most of the information known about the Soviet activities has come from two men: Ken Alibek and Gennadi Lepyoshkin, who were both scientists who worked on the island before defecting to the US. The two defectors openly admitted the activities that they were involved in, many of which included testing deadly viruses and super-strains on various animals. The tests were conducted under the cover of darkness to prevent being detected by satellites. Aware that any outbreaks on the Soviet population could arouse the suspicion of the West, the Soviets did attempt to contain the effects of their tests. Recognising that super-strains could be transferred to the mainland by insects or birds, the scientists reportedly poisoned the whole testing area to kill off all living species. In the Soviet war on nature, nothing was sacred.

Gennadi Lepyoshkin also stated that the Soviet regime realised the benefits of biological weapons as opposed to developing nuclear capabilities: it calculated that to destroy one square kilometre it would cost $2,000 with conventional weapons, $800 with a nuclear weapon, $600 with a chemical weapon and only $1 with a bio-weapon. After discovering how cost-effective the weapons could be, the Soviet scientists refocused their attention on creating the destructive super-strains, disregarding the potentially devastating impacts of testing on the island.

Lepyoshkin was a supervisor of scientific teams on Vozrozhdeniye Island in the 1970s, but he told the Americans that all the work was carried out in a very relaxed environment. He recalls the still fresh and deep Aral in which they swam during their leisure time. 'The island was beautiful,' Lepyoshkin claims. He recalls how one female scientist dropped a Petri dish of anthrax and tried to cover it up, and was not even punished for her action. 'Nobody got sick,' he says, 'so nobody bothered.' Lepyoshkin also claimed that not all the work he did was negative. 'We discovered new methods to improve the immune system,' he says. 'We developed an anthrax vaccine that was given to the whole army, and it's considered to be the best in the world. Same with our plague vaccine; it's been used for more than 40 years.'

The other defector, Ken Alibek, was the former head of the Soviet germ warfare programme and also worked on Vozrozhdeniye Island. He openly admitted to the Americans the various different super-strains of bacteria they worked on, resistant to all known antibiotics. He claims to have been ordered to prepare a strain of anthrax, smallpox and bubonic plague to be put in a warhead aimed at the United States, targeted on New York, Boston and Chicago. Alibek openly admitted that in a certain scenario the devastation would be catastrophic. Animal testing was common on the island, he confirmed, utilising guinea pigs, hamsters, rabbits, cows, horses, donkeys and even monkeys.

According to Alibek, the story of Vozrozhdeniye Island gets worse. He says, 'A train was heading towards the Aral Sea carrying 100 tonnes of anthrax from Stepnogorsk. There were orders straight from Moscow to bury the anthrax on Vozrozhdeniye and never to speak of it again. Covered in bleach, the anthrax was shipped to the island in steel barrels. Officials decided instead of burying them in the barrels they would just dump them in pits and pour a little more bleach on top of them just to be sure!'

'I knew the weapons would never be used,' Alibek says. 'When nuclear weapons were made, no one thought they would be used. You'd have to be mad to use them. But now that there's terrorism, it's more scary. You know, biological weapons are cheap.'

In 1991, the island was abandoned altogether and remains to this day one of the most hazardous places on the planet.

Decontaminating Vozrozhdeniye

Unique environmental, social, health and security concerns have been created by weapons testing on Vozrozhdeniye, and these concerns are heightened by the continuing desiccation of the Aral Sea. Since the Aral began to shrink in the 1960s, the island has grown from around 180 to 1,800 square kilometres. The southern part of Vozrozhdeniye Island became connected to the mainland in 2001. In other words, Vozrozhdeniye stopped being an island.

Fears persist that strains of virulent diseases may still be viable at the former test site. The chief concern is that some weaponised organisms may have survived and could escape to the mainland via infected rodents, looters or terrorists that might gain access to them.

The extent of decontamination and the current state of the island is difficult to ascertain. Allegedly, Soviet scientists took steps to decontaminate the island before departing, but this has never been confirmed. Similarly, Moscow refused to provide Uzbekistan with any information relating to the tests and, as a result, the Uzbek authorities signed a bilateral agreement with the USA in May 1995.

That year, Uzbek officials invited experts from the Pentagon to analyse 11 pits where anthrax had been buried. A second visit in October 1998 discovered that live organisms, including live anthrax spores, could still be found in six of the pits. Although details were not revealed, electric power was suggested to heat the soils and kill the bacteria. Kazakh scientists have not carried out any investigations into this problem in the part of Vozrozhdeniye Island that lies in their territory, due to a lack of funding.

Heightened global security tensions led to further US involvement in the new millennium, and the US government has both donated money and sent experts to help Uzbekistan ensure the destruction of any surviving weaponised pathogens. In October 2001, the US Department of Defense and the Uzbek Defence Ministry signed an agreement allowing the Cooperative Threat Reduction programme to spend up to $6 million to destroy residual spores, and therefore reduce risks to the environment and human health. In 2002, an expedition was led by officials from the Pentagon's Defense Threat Reduction Agency to neutralise what is probably the world's largest anthrax dumping ground. The team of 113 people neutralised between 100 and 200 tonnes of anthrax over a three-month period.

Kazakh and foreign oil companies approached the Anti-Plague Institute in Almaty last year to investigate the possibility of exploration. The institute, which now studies a range of infections, strongly advised against it. 'Without research about the danger or safety of visiting the island, we cannot give any guarantees,' said Bakyt Atshabar, its director.

Nevertheless, on the Uzbek side of the former island, oil exploration is already underway, near to some of the trenches dug by Soviet soldiers where large quantities of deadly bacterial sludge were poured into the earth and sprayed with bleach. Only time will tell if this hasty decontamination process was effective.

Uzbekistan has inherited a global problem with the desiccation of the Aral Sea and the subsequent exposure of Vozrozhdeniye Island, and it pins the blame firmly on its upstream neighbours, such as Tajikistan and Kyrgyzstan. The Uzbeks claim these countries are hoarding water and experimenting with dangerous hydroelectric power projects whose construction began during the Soviet era. Water lies at the heart of the simmering tensions that define inter-state relations in Central Asia and are a lasting legacy of Stalin's war against nature.

CHAPTER TEN

Water Is Life

Tajikistan and the Nurek and Rogun Dams

The renowned American ecologist Rachel Carson once said: 'In an age when man has forgotten his origins and is blind even to his most essential needs for survival, water, along with other resources, has become the victim of his indifference.'

Water is essential to all human and other life forms. A shortage of water can have devastating consequences, causing a sharp fall in food production and threatening life itself. Water shortages can lead to huge movements of people, and can even be the root cause of conflict. But to find water disputes in a land of plenty is both bizarre and unusual.

Nowhere is this more evident than in the Ferghana Valley. During Soviet times, everything was planned centrally by Moscow. In his ongoing war to 'tame' nature, Stalin ordered great reservoirs to be constructed in the mountainous territories of Tajikistan and Kyrgyzstan to supply water for agricultural irrigation in the downstream countries during the hot summer months. In return, some of the rich coal and oil reserves of the downstream republics were sent to Tajikistan and Kyrgyzstan to provide them with energy in the freezing snows of winter.

In the mid-1920s, Stalin cynically exploited the water situation in Central Asia to help the development of his policy of divide and rule. The Soviets created two small republics, Tajikistan and Kyrgyzstan, endowed with enormous water resources, although little in the way of agricultural land, and three large republics, Turkmenistan, Uzbekistan and

157

Kazakhstan, with huge agricultural potential but virtually no indigenous water supply.

Professor Sara O'Hara of the University of Nottingham is a world-renowned expert on the Aral Sea and has visited the area many times. She attended a conference on the Aral Sea that I organised in the European Parliament in Brussels in late 2010.

Dr O'Hara says: 'In effect, the Soviet administration created a situation which would ensure competition between water-surplus and water-deficit republics. This situation worked to Moscow's advantage in two ways. First, disputes over water reinforced the national distinctiveness of the republics, thus limiting the potential for regional co-operation which would threaten Soviet control. Second, as competition for water increased, the republics were forced to ask Moscow to intervene, a role it was more than willing to undertake. In short, water policy was central to Moscow's efforts to divide and rule the region.'

During the Soviet era, Moscow put pressure on Kyrgyzstan and Tajikistan to empty their reservoirs during the summer months so that the cotton fields in downstream states could be irrigated. All of this changed after independence. Kyrgyzstan and Tajikistan now store the water during the summer and release it during the winter to generate hydroelectric power. This has led to a number of serious disputes since independence. In 1993, in 1998 and in 2001, Kyrgyzstan was blamed for releasing too much water from the Toktogul Dam down the Syr Darya River during the winter and not enough during the summer. The result was that a lot of cotton fields were flooded in Uzbekistan and Kazakhstan. The same kind of problem is also occurring in the Amu Darya river basin.

As a direct response to these problems, the downstream nations like Turkmenistan and Uzbekistan began the feverish construction of reservoirs, intent of maintaining their own summer water supplies, using the water for irrigating great swathes of cotton and even rice. This activity has contributed to the worsening desiccation of the Aral Sea.

Meanwhile tensions between the Central Asian neighbours flared. Uzbekistan closed its borders with Tajikistan and withdrew unilaterally from the Central Asian electricity grid, which had provided a free flow of energy between the nations since Soviet times. The borders between these two republics often remain closed for long periods, with a consequent huge back-up of trucks, goods vehicles and cargo trains on either side, creating havoc for traders and damaging the fragile economies of both nations.

The root of the problem, according to Uzbekistan, is the plan to build a new giant hydroelectric power plant in Tajikistan. The Rogun Reservoir on the Vakhsh River, a tributary of the Syr Darya, has become a focus of controversy. When operational, Rogun will produce a massive 3,600MW at peak capacity. Over 45km of underground tunnels had already been completed during the Soviet era. These are now being refurbished by 7,000 workers, who are engaged day and night in the construction of a giant underground hall where the turbines will be located. It is intended to dam the Vakhsh River in the steep, narrow valley in which Rogun is located with a towering 335m-high stone-and-clay embankment.

The Uzbeks in particular are deeply alarmed that this dam would be vulnerable to severe seismic activity; if ever breached, they say it would have catastrophic consequences for down-stream countries like theirs. German and Pakistani experts have been employed to assess the Rogun project by the World Bank. A final assessment was originally supposed to be pub-lished in 2011, but the complexity and sensitivity of the decision has caused significant delays. However, it seems likely the Dam will get the green light. There is a precedent, after all, in the nearby Nurek hydropower station, also on the Vakhsh River. Nurek has functioned successfully for 40 years without any problems, despite constant earthquakes in the area. Like the proposed development at Rogun, the Nurek dam is over 300 metres in height, and is also built of rocks and clay and designed to withstand severe earthquakes.

Opponents of the Rogun hydro project often cite the ex-ample of the Sayano–Shushenskaya Dam, located on the Yenisei River near Sayanogorsk in Khakassia, in south-central Siberia. It is the largest power plant in Russia and the sixth-largest hydroelectric plant in the world.

The plant is operated by RusHydro. Like Nurek and the proposed dam at Rogun, the Sayano–Shushenskaya Dam was designed to withstand earthquakes of up to 8 on the Richter scale, and was recorded by the *Guinness Book of World Records* as the strongest construction of its type. The dam supports the Sayano–Shushenskoe Reservoir, with a total capacity of 31.34 cubic kilometres and a surface area of 621 square kilometres.

A series of accidents have taken place at the Sayano–Shushenskaya Dam, the most devastating of which occurred on 17 August 2009, when turbine 2 of the hydroelectric power station exploded. The turbine hall and engine room were flooded, the ceiling of the turbine hall collapsed, nine of the ten turbines were damaged or destroyed, and 75 people were killed. The entire plant output, totalling 6,400MW, and a significant portion of the supply to the local grid, was lost, leading to widespread power blackouts in the local area and forcing all major users such as local aluminium smelters to switch to diesel generators.

There had been a history of problems with the number two turbine prior to the 2009 accident. In fact, reports of problems with this turbine date as far back as 1979. Between 1980 and 1983, many more problems occurred with seals, turbine shaft vibrations and bearings, so that by the year 2000 it was decided that a complete reconditioning of Turbine 2 should be carried out. When the turbine was dismantled, cavities up to 12mm deep and cracks up to 130mm long were found on the main wheel and repaired. Many other defects were found in the turbine bearings and subsequently repaired. In 2005, further repairs were made to Turbine 2.

From January to March 2009, Turbine 2 was again undergoing scheduled repairs and modernisation. During the course of the repair, the turbine blades were welded. However, when the turbine was refitted on completion of the repairs, the wheel was not properly rebalanced, leading to increased vibration and more pressure on the main bearing. This was logged, but no remedial action was taken, with catastrophic results.

On the night of 16–17 August, the level of vibration in the

badly fitted wheel increased substantially. On the morning of 17 August 2009, around 50 people were working in close proximity to Turbine 2. The general director of the plant was celebrating his seventeenth anniversary as boss, and in the early morning he drove to nearby Abakan to collect some guests who had been invited to his celebration party. In his absence, and despite increasing fears over the level of vibration emanating from the number two turbine, none of the remaining workers felt they had the authority to take decisions affecting the operation of the plant. It cost many of them their lives.

At 8.13 a.m. local time, there was a loud explosion in Turbine 2. The turbine cover shot up, and the 920-tonne rotor careered out of its seat. Water gushed under huge pressure from the cavity of the turbine into the machinery hall, flooding the entire area and all of the rooms below, drowning dozens of workers. At the same time, an alarm was received at the power station's main control panel and the power output fell to zero, resulting in a local blackout over the whole region. The steel gates to the water intake pipes of the turbines, weighing 150 tonnes each, were closed manually.

The explosion of Turbine 2 also affected turbines 7 and 9, effectively destroying them both. The explosion also caused roofs and ceilings to collapse, damaging turbines 1, 3, 4, 5, 8 and 10. Only Turbine 6 survived, as it was being repaired at the time. The flooding of the turbine room also caused severe electrical scarring to the transformers. Transformers 1 and 2 were destroyed, while transformers 3, 4, and 5 were left relatively unscathed.

Power generation from the station stopped completely after the explosion, causing widespread blackouts in residential areas, which received power after electrical power was diverted from other plants. Power to certain areas was not completely restored until 19 August 2009. Aluminium smelters in Sayanogorsk and Khakassia were completely isolated from the grid, and power supplies had to be replaced using alternative energy sources.

The incident also created a severe oil spill, releasing upwards of 40 tonnes of transformer oil into local watercourses. This

severely affected the local environment, and traces of oil were
recorded as far as 80km downstream of Yenisei, killing 400
tonnes of cultivated trout in two riverside fisheries. By 19
August 2009, the 15km-long spill had reached Ust-Abakan,
where it was cordoned off with floating barriers and chemical
absorbents. The oil spill took over two weeks to properly
clean up.

It is the environmental impacts and the effects on the local
and regional power supply that worry Central Asian states.
Moreover, some representatives of Central Asian countries are
fearful that if such an accident were to happen in an upstream
area in Tajikistan or Kyrgyzstan, then there would be disastrous flooding downstream, in addition to environmental
devastation and loss of power supply. These are the reasons
why the Sayano–Shushenskaya accident is commonly refer-
enced in Central Asian water-management disputes.

Ninety-five per cent of electricity in Tajikistan is generated
from hydroelectric power plants. The overhead lines and
infrastructure needed to sell this electricity to neighbouring
countries like Afghanistan and Pakistan are already being built.
The new reservoirs being planned at Rogun will take ten years
to complete before they are fully operational, but the Tajiks
guarantee that they will continue to supply their downstream
neighbours with the same amount of water that they enjoy now
via the Amu Darya and Syr Darya rivers. They claim that their
new system of dams will provide a properly managed water
source which will benefit everyone. The Tajiks point out that
60 per cent of the rivers which serve Central Asia are sourced in
Tajikistan. They claim that they have never and will never
restrict water flow to their downstream neighbours.

Once it is operational, Rogun will provide a source of green,
environmentally friendly energy which is both cheap and
plentiful, and will meet the needs of Tajikistan while also
providing essential energy for neighbouring countries like
Afghanistan and Pakistan. It seems like a win-win situation
in an area that is desperate for energy. Combined with a more
strategic use of water in the downstream countries, with
concrete-lined reservoirs and droplet irrigation, there is no

reason why the abundant water resources of Central Asia cannot be distributed fairly and used in a way that enhances, rather than threatens, the future of the Ferghana Valley.

I visited Tajikistan in September 2010 and discussed these controversial issues with the Tajik minister for energy and industry and the speaker of the Majilis. I also met the chairman of the committee on environmental protection and the first deputy minister of melioration and water resources. They took me to visit the impressive Nurek and Rogun hydroelectric power projects on the Vakhsh River.

Tajikistan has emerged after a bloody civil war following independence as a stable but relatively poor country. It has great potential to develop mineral and water resources, provided it can achieve significant levels of inward investment. But it is Tajikistan's geographical position, nestling in the high Pamir Mountains on the rooftop of the world, with Afghanistan, Iran and the other Central Asian republics as neighbours, that make it strategically important for the West. This is a highly sensitive area.

War is still raging in neighbouring Afghanistan, where the Taleban are continually active. Islamic terrorists lurk in the mountains in nearby Pakistan. Iran seeks to spread its ideological influence across the whole region. Drug trafficking is rife. But Tajikistan, under the authoritarian leadership of President Emomali Rahmon, stands guard. President Rahmon claims to be at the forefront of the fight against drugs and terrorists and, as such, says he is a key strategic ally for the West.

His name was originally Rahmonov until 2007, when he ordered his countrymen to drop Russian-style surnames, in a break with the nation's Soviet past. Rahmon was elected president in 1994 and re-elected in 1999, when his term was extended to seven years. In 2006, he won a third term in office in an election which international observers said was neither free nor fair. Opposition parties boycotted the vote, dismissing it as a Soviet-style staged attempt at democracy.

The president has a firm grip on power. His People's Democratic Party holds virtually all the seats in parliament. Western observers said legislative elections in 2005 and 2010 failed to meet international standards, but Mr Rahmon does retain substantial public support. Tajikistan is still very poor, but many people are thankful they no longer have to face the civil war of the 1990s. President Rahmon emerged the victor from that vicious conflict, perhaps providing him with more legitimacy as a ruler than many of his neighbours in Central Asia.

Tajikistan also has a major role to play in bringing stability to this volatile region. Through the exploitation of its massive water resources, it will soon be able to meet all of its own energy requirements, while at the same time exporting electricity to its neighbours.

Driving out of Dushanbe on the way to the great Nurek hydropower plant, it quickly became evident how poor Tajikistan really is. The roads, which were well paved in the city centre, suddenly deteriorated in the suburbs into potholed and broken tracks. It was impossible to travel at more than a crawl, as we dodged newly dug trenches and deeply gouged holes. Cars coming in the opposite direction were executing similar manoeuvres, so the entire road resembled the dodgems at a village fair.

Tajikistan is a rugged, mountainous country, with lush valleys to the south and north. Tajiks are the country's largest ethnic group, with Uzbeks making up a quarter of the population, over half of which is employed in agriculture and just one-fifth in industry. Nearly half of Tajikistan's population is under 14 years of age.

Tajikistan is Central Asia's poorest nation. The five-year civil war between the Moscow-backed government and the Islamic fundamentalist opposition, in which up to 50,000 people were killed and over one-tenth of the population fled the country, ended in 1997, following a United Nations-brokered peace agreement. The population is only seven million, but the country's economy has never really recovered from the civil war and poverty is widespread. Almost half of Tajikistan's GDP is earned by migrants working abroad, especially in Russia, but the recession in 2009 threatened that income. The country is also dependent on oil and gas imports.

Tajikistan has relied heavily on Russian assistance to counter continuing security and economic problems.

Skirmishes with drug smugglers crossing illegally from Afghanistan occur regularly, as Tajikistan is the first stop on the drugs route from there to Russia and the West. Tajikistan has also been accused by its neighbours of tolerating the presence of training camps for Islamist rebels on its territory, an accusation which it has strongly denied.

In October 2004, Russia formally opened a military base in Dushanbe, where several thousand troops are stationed. It also took back control over a former Soviet space monitoring centre at Nurek. These developments were widely seen as a sign of Russia's wish to counter increased US influence in Central Asia.

The mountainous landscape of Tajikistan means that scarce water resources and limited water availability are not a concern. However, as an upstream country, Tajikistan has responsibilities towards downstream states, and their water usage has serious impacts on lower level countries. Water pollution is one issue that has gained a trans-boundary dimension and now affects relations between Tajikistan and Uzbekistan. In intra-state terms, the quality of drinking water is a major problem due to poor sewage treatment and informal garbage dumps causing contamination, and water-borne diseases such as typhoid and cholera.

Due to a very high dependency on hydropower and agriculture, Tajikistan is highly susceptible to natural disasters. Even when I was there, several minor earthquakes occurred. On the evening I departed, as I waited in the airport lounge at Dushanbe, a tremor left the floor rippling and the chandeliers swinging from side to side. A second tremor caused some of the airport staff to run for shelter under doorways and outside the building, although others nonchalantly said that this was routine and nothing to worry about.

Land degradation is another current issue in Tajikistan, as is deforestation, desertification and loss of biodiversity. Desertification is estimated to affect approximately 60 per cent of irrigated land, and salinisation, caused by poor irrigation practices, has become a widespread problem.

On my first visit to Tajikistan, I was taken to see both the existing Nurek hydro project and the controversial Rogun Dam. After several hours driving on broken roads, we finally arrived at the town of Nurek, specially built by the Soviets to house engineers and workers from the Nurek hydroelectric plant on the Vakhsh River. We made our way past a large and incongruous plinth on which was mounted a gigantic yellow-and-red, caterpillar-style tunnelling machine, looking, for all the world, like some scary space monster from a sci-fi movie. We drove through the town and on up to the dam, where we were welcomed by the plant director, Rustam Zogakov. Mr Zogakov explained that he had started work at Nurek as an apprentice engineer during the Soviet times and had risen to become the chief engineer and now the boss. He said he was also a senator in the Tajik senate in Dushanbe.

Mr Zogakov took us in a creaking lift up to the viewing platform in the giant turbine hall. Here we could see the nine turbines that produce more than 3,000MW of electricity. Far below, great plumes of water gushed from overflow pipes, thundering into the turquoise water of the river. The resulting towering cloud of spray they create apparently keeps local temperatures one or two degrees cooler in the heat of summer than elsewhere in the region. A large tunnel to one aside of the river had the legend 'Water Is Life' painted in English over its arch. We were shown around the control room and then taken back down to the car park. Here we we were ushered into a four-wheel drive Lada and driven up a winding track until we reached the very top of the dam. On one side, a startlingly blue lake stretched away as far as the eye could see, perfectly still and mirror-like. Mr Zogakov said the 300-metre-high stone-and-clay embankment had created a 70-kilometre-long reservoir. On the other side, the wall of the dam cascaded down into the narrow valley far below. The turbine hall and hydro plant from this perspective looked like children's toys.

'The Soviets always overengineered everything,' Mr Zogakov said. 'They knew the dangers of earthquakes in this zone and they designed Nurek accordingly. It was built to last.' Mr Zogakov pointed down towards the buildings far below. 'The dam was constructed between 1961 and 1980. It has a central core of concrete forming a solid barrier within a 300-metre-high rock and earth-fill embankment, making it the tallest dam in the world. Only Rogun, further up this same Vakhsh River, will be taller when it is eventually completed.'

He moved across the viewing area to a point where we could get a clearer view of the massive embankment. The sun was beating down on us at this high point, lighting up the almost vertical sides of the ravine, where the Vakhsh River had been halted by this incredible feat of Soviet engineering more than 40 years ago. This really was a symbol of the Soviet Union's war on – and in this case, victory over – nature. He gestured for me to follow. 'The volume of this embankment mound is 54 million cubic metres,' he said, pointing to the huge wall. 'As you have seen, the dam includes nine hydroelectric Francis turbines. The first was commissioned in 1972 and the last in 1979. The reservoir that you can see here fuels the hydro-electric plant, but the stored water is also used for irrigation on local agricultural land. Irrigation water is transported 14 kilo-metres through the Dangara irrigation tunnel and is used to irrigate about 700 square kilometres of farmland.'

I quizzed Mr Zogakov about the infamous accident at the Sayano–Shushenskaya Dam in Khakassia, Russia, in 2009. He dismissed any alleged connection with Nurek out of hand. 'These are totally different structures,' he said. 'Nothing like that could ever happen here.' However, I did notice as we descended back down to the riverside that a huge turbine was undergoing refurbishment on a concrete platform behind the turbine hall. Deep cracks in its blades were being welded and polished, presumably in a routine maintenance operation.

At the riverside, a large table had been set on a platform next to the swirling waters of the Vakhsh River. The table was groaning with food, and the senior engineers from the plant had gathered to join their boss and me for a lavish lunch. The hospitality of the Tajiks is legendary, and in due course the inevitable sheep's head arrived and was proffered to me to do the ceremonial carving. I think the sheep in Central Asia must get nervous every time they hear I am coming! Several vodka toasts were exchanged and then we said our farewells and set off for the other big hydro project in Tajikistan – Rogun.

Another three hours in the car bumping over tortuous roads brought us to the town of Rogun, specially constructed in the 1970s to house thousands of Soviet workers and engineers whose job it was to construct the world's highest dam. The Rogun hydroelectric project is located further up the Vakhsh River from Nurek. Construction began in 1976 but stopped abruptly with the collapse of the Soviet Union. In February 2007, Russia announced a partnership with Tajikistan to complete the dam, but later this was abandoned because of disagreements over who had the controlling stake in the project. The Tajik government then set about trying to raise US $1.4 billion to complete the project, however, by April 2010, it had only succeeded in raising $184 million, mostly through selling shares to the impoverished Tajik public – enough to continue construction work for a further two years.

Forty-five kilometres of underground tunnels built by the Soviets are now being refurbished by 7,000 Tajik engineers and builders. Huge dumper trucks are hauling tonnes of rubble and clay out of the tunnels, which have been bored deep into the steep walls of the ravine.

It appears that, despite the controversy created by this project and the anger of the downstream Uzbeks, the Tajiks are determined to complete all parts of the hydro plant apart from the construction of the dam itself. The actual damming of the Vakhsh River will have to await the verdict of the two World Bank experts who were appointed to assess the safety and viability of Rogun. Uzbek President Islam Karimov has called the Rogun hydropower plant a 'stupid project'.

I was driven deep underground in a four-wheel-drive vehicle, splashing through mud and large pools of water in the dimly lit network of tunnels. Occasionally we would screech to a halt as a dumper truck loomed out of the tunnel mouth directly ahead of us, necessitating much reversing and cursing by our driver. Finally, we emerged into a monstrous underground cavern carved out of the mountain. This was the turbine hall, our guide explained. 'This is where six turbines with a capacity of

3,600MW will be located,' he said. Without offering hard hats or any protective clothing, we were invited to trip precariously along piles of steel rods and other building material, while the clatter and clanging of construction work went on around us.

In the glow of several dozen arc lights, I stared around at this bizarre scene. It reminded me of a James Bond movie. The underground hall could easily house an entire football pitch and still leave room to spare. There was a mass of activity underway. Sparks from electric arc welders lit dark corners of the cavern. Water dripped and ran in rivulets down the steep rocky sides. Trucks and drills jostled for space. Recalling that we were 50 metres underground in the middle of a seismic zone, I was greatly relieved when our guide signalled it was time to head back to the surface. Breathing in the cool mountain air once again back on the surface, I glanced at a huge poster of President Rahmon that dominated the main car park at the Rogun plant. His beaming face and open arms gestured to a future filled with promise.

I returned to Tajikistan in May 2011. After my previous visit, I had written a supportive article in the press about the Rogun hydro project and now President Emomali Rahmon himself wanted to meet me. Arriving on a flight from Istanbul, I was collected in the small hours of the morning in the VIP lounge at Dushanbe international airport and whisked off to my hotel in a Ministry of Foreign Affairs car. I managed to catch around three hours' sleep before being picked up again and driven to the grandiose presidential palace. I was accompanied once again by my trusted interpreter and adviser Dr Kamila Magzieva. However, when we arrived at the presidential palace, we were met by the foreign minister, Hamrokhan Zarifi, who said that Kamila would have to wait with him in the foyer, as I had a private audience alone with President Rahmon.

In the event, Tajik not being one of my strong points, when I was ushered into a vast, ornate chamber to be shaken warmly by the hand by President Rahmon, I was relieved to see that his own interpreter was also there. I was shown to an elaborate golden throne next to an even more elaborate one occupied by the president. The interpreter pulled up a chair facing us. The president began by thanking me for my 'excellent' newspaper article about Rogun. He explained the importance of the project for Tajikistan and said that it was incomprehensible to him why Uzbekistan's president was so opposed to it.

He then launched into a long and at times lurid account of his campaign against terrorists, at one point rolling up his trouser leg to show me scars he had received when a hand grenade had been tossed into a crowd of students he was addressing, killing and wounding several. He said his country was constantly engaged in trying to stop the infiltration of Islamic terrorists and drug runners from Afghanistan and Pakistan, and he fervently hoped that the West would recognise the strategic importance of supporting Tajikistan in this battle.

President Rahmon suddenly lent forward and grabbed me tightly by the wrist. His face was only a few inches from mine. 'As you know, I am coming to Strasbourg next week for meetings with the president of the European Parliament and for a debate with members of the Foreign Affairs Committee. I hope that I can meet you there, and I hope that you will repeat your support for our Rogun hydro project.'

Several times, the huge carved wooden doors to the president's salon swung silently open and his senior officials – and, at one point, even Foreign Minister Zarifi himself – peered in to see why our meeting, which had been scheduled to last only 30 minutes, had continued for over one and a half hours. Each time, President Rahmon waved them away with a flick of his hand. Finally, when I had briefed him fully on what to expect from his meetings in Strasbourg, he summoned the others to join us. A lengthy discussion ensued, which neither Kamila, who speaks Russian and Kazakh, but not Tajik, nor I could understand. Then Mr Zarifi explained. He said that President Rahmon was due to make one of his regular visits to the hydro project at Rogun the next morning and wanted me to join him. We were to meet at Dushanbe airport at 6 a.m. to join reporters, TV crews and government ministers on a fleet of helicopters which would fly us to Rogun.

The next morning we were up at the crack of dawn and driven to the airport by our Ministry of Foreign Affairs minder.

Mr Zarifi was already there. Promptly, at 6 a.m., we were taken across the tarmac to three large military helicopters. There were already two camera crews and four or five other officials and reporters in the one I was directed to. No sooner had we clambered on board than the rotors started to turn and we were thundering across the mountains towards the valley of the Vakhsh River.

Tajikistan supplies most of the water consumed by Uzbekistan, Turkmenistan and Kazakhstan from more than 900 fast-flowing rivers, springing from glaciers high in the Pamir Mountains. That its water spells opportunity is in little doubt. But Tajikistan, Central Asia's poorest state, faces some tough questions over how best to balance its own needs with the rest of the region. Energy is in critically short supply, with some 40 per cent of the country's electricity absorbed by the aluminium industry alone. Without new energy supplies, the furnaces and lights could go out altogether.

President Rahmon is determined to show that his country doesn't need a planned Soviet-style economy to pull itself into the twenty-first century. Soon Rogun's reservoir will hold over 13.4 square kilometres of water at an average depth of 400 metres, flooding this mountainous valley and providing enough energy to meet all of Tajikistan's needs, with enough left over to sell to neighbouring Afghanistan and, beyond, to Pakistan.

A secure supply of electricity from Tajikistan will transform the economies of these ravaged regions and provide new sources of employment and opportunities for their impoverished citizens. Yet the Soviet command economy had one key advantage. It could keep rivalries between its respective republics in check, by force if necessary. These days the geopolitical situation in Central Asia is more complex. Competing with the basic economic needs of Tajikistan are the loud claims of neighbouring Uzbekistan that the new dam is at risk from earthquakes, or that it will allow the Tajiks to control the flow of water to its people. If the World Bank gives the Rogun Dam a clean bill of health, the West, and the EU in particular, must endorse its opinion for the sake of Tajikistan's and Central

Asia's long-term stability. The alternatives – long-term poverty and desperation for its long-suffering people – are in no one's interests.

Our helicopter hovered above the helipad next to the town of Rogun, specially constructed and recently enlarged to accommodate the 7,000 workers and their families. Mr Zarifi shouted to the pilot and gesticulated with his hands, pointing northwards. The pilot banked steeply and we were away again, racing up the valley of the Vakhsh, the steep walls of the ravine rising perilously close to our swirling rotors. Mr Zarifi explained to me that we were meant to land further upstream, at the site of the gigantic reservoir, where we would again meet up with President Rahmon.

Soon we could see the bed of the reservoir spread out beneath us, swarming with men and machines. Mechanical diggers were filling endless rows of dumper trucks with tonnes of earth and rocks, deepening the bed of what would soon become an enormous inland lake. We touched down on a grassy strip, where a party of local officials stood waiting. I stared around at the encircling hills, realising that I was standing on what will soon become the bed of a lake, on fields and rocks never again to be seen by mankind, perhaps for eternity. I was introduced to the site manager and senior engineer, and to several members of the Tajik cabinet who had assembled for the tour of inspection. Kamila and I were kitted out with steel-toed wellington boots and hard hats as we waited for the arrival of President Rahmon's helicopter.

Shortly afterwards, his yellow chopper swooped over the mountains and settled next to ours in the grassy field. The president and his entourage emerged. He made straight for me and gave me a huge bear hug, advising everyone that I was a 'leading expert from the European Parliament', here to see for myself the reality of Rogun. We were ushered towards a fleet of four-wheel-drive vehicles and soon were roaring in a cloud of dust towards the first of the major underground tunnels which honeycomb the mountains around and under Rogun.

As we plunged deeper and deeper underground, the scale of this project became more and more obvious. Just as I had seen on my first visit to Rogun, almost nine months earlier, everywhere vast caverns, first carved by the Soviets, are being refurbished and reinforced. Every so often we would stop and the president would grab my wrist and guide me to a feature which he would then describe in detail. He pointed to the tens of thousands of holes being drilled into the ceilings, walls and floors of every tunnel so that steel cables and concrete can be sealed into them, strengthening the overall structure. He took time to explain to me complex mechanical designs on specially provided plans. He led me to the gaping chasm, hundreds of metres underground, where the giant turbines will be installed. At every stop, the president greeted the workers, shook their hands and spoke to them for minutes at a time, missing no one. It was obvious from the looks on their faces that they greatly appreciated their president's friendship and willingness to join them in their underground labours.

Eventually, after around four hours, we emerged once more into the sunlight, mud-spattered and weary, but well briefed on the intricacies of this enormous project. A new bridge had just been completed, crossing the raging torrent of the Vakhsh, and President Rahmon had to cut the tape and pronounce the construction officially open. He beckoned me to join him and, as a large crowd of workers and local dignitaries cheered, he cut the tape and we walked across the bridge, shaking hands and chatting to workers on the way.

President Rahmon said that it was time now for him to resume his official site visit but that he had instructed Foreign Minister Zarifi to take me for lunch higher up in the Pamir Mountains. We said our goodbyes and, together with Kamila and Hamrokhan Zarifi, I was driven back to Rogun town, where a smaller, white civilian helicopter was waiting with three smartly uniformed crew. It transpired that this was one of a fleet of choppers belonging to the Aga Khan, who has major business interests in Tajikistan and is a personal friend of Mr Zarifi. He allows the foreign minister to use one of his helicopters whenever he needs it, and we were introduced to the senior pilot, a German, and his two Tajik co-pilots, as we were strapped into comfortable white leather seats.

Once more we roared off into the mountains and this time made our way for more than a hundred miles until we reached a dry riverbed, which Mr Zarifi claimed was the spot he was looking for. We landed at the foot of a spectacular waterfall, and some local villagers led us up more than 200 steps carved out of the cliff face until we reached the top of the waterfall, where a large canopied platform had been built precariously overhanging the foaming water. The platform was covered in rugs and cushions, and a low table in the centre was groaning under the weight of dishes of fruit, vegetables and meats.

We clambered onto the platform and made ourselves comfortable on the plush cushions. Villagers, laden with dishes of steaming delicacies, were racing up and down the steps from a small hut far below, where the cooking seemed to be taking place. There were bowls of nuts and wild white brambles, collected from the local forest. There were jars of local forest honey and hot, oven-baked chapattis. I was offered a plate of roasted birds, which, I was informed, were parrots from the mountain forests. I declined.

Mr Zarifi gestured to the array of bottles adorning the table and asked me what I would like to drink. He suggested some vodka, cognac or whisky. I said that it was far too early in the day for strong spirits, but if there was any wine available I would welcome a small glass. Mr Zarifi stretched across the table and grabbed two of the wine bottles which had been brought for our picnic lunch. He quickly put them down again with a grunt of dismay. 'Georgian wine, and too sweet,' he grumbled. He summoned his assistant and ordered him to fetch his briefcase. The poor chap set off at high speed down the 200 steps to the helicopter and about ten minutes later returned, red-faced and panting with exertion. He handed the briefcase to the foreign minister, who opened it and pulled out a magnificent bottle of Chateau Margaux 1990, one of the finest vintages from Bordeaux in the last half century. 'I thought you might like a glass of wine with lunch,' Mr Zarifi exclaimed, 'so I grabbed this from the foreign ministry cellar on the way here.' He uncorked the bottle, ceremoniously smelled the cork, and poured Kamila, me and himself a generous glass.

I can honestly say that sitting at the top of a raging waterfall high in the Pamir Mountains, eating wild white brambles with honey and chapattis and sipping Chateau Margaux 1990 must be one of the finest culinary experiences I have ever had. Sadly, this idyll was not destined to last. A loud roar of thunder rumbled across the darkening sky, followed by jagged flashes of forked lightning. Within minutes, a torrent of rain began to fall. Our German pilot came panting up the slippery steps saying we must leave immediately. 'Helicopters can handle heavy rain,' he said, 'but thunder and lightning is another matter.'

I gulped the last of my glass of Margaux and left the shelter of our canopied platform. Within seconds, as I slipped and slid down the wet steps, I was soaked to the skin. We clambered on board the helicopter, which quickly steamed up from our wet clothes. The co-pilot slammed the door shut and the rotor blades began to turn. As we rose into the air, a great flash of lightning and crack of thunder sent shudders through the entire craft. The pilot wheeled to one side and we roared off under the heavy storm clouds, heading for the Afghan border, where Mr Zarifi was keen to show me a new bridge across the Panj River dividing Tajikistan from Afghanistan. The bridge had been paid for by the US, and Mr Zarifi wanted me to meet the border guards on either side of the river to see how effective they are in combating drug trafficking and terrorists.

Within ten minutes we had emerged from the storm and were now flying across mountainous valleys and deep gorges in bright sunshine. But suddenly the chopper wheeled again and set off in a new direction. The co-pilot removed his headphones and shouted to us that the Afghan border area was also stormbound, and even Dushanbe airport had been closed to helicopter traffic because of the weather, so we had no alternative but to return to Rogun and wait for the storm to pass.

Back at Rogun, we landed next to the military helicopters which had brought us there earlier that morning and were now also grounded because of the weather. The president had apparently headed back to Dushanbe by car, unable to wait any longer for the weather to improve. We hung around for a further two hours, while our German pilot kept in constant contact with Dushanbe air-traffic control by radio. Finally, around five o'clock in the evening, we got the all clear and were able to make our final journey back to the capital.

My attempt to visit Afghanistan had, for the present, been put on hold, but Mr Zarifi was adamant that I must return soon to Tajikistan so that he could take me to the Afghan border and introduce me to his country's most volatile neighbours.

CHAPTER ELEVEN

Afghanistan

Sitting on the very roof of the world, war-torn and beleaguered Afghanistan can only dream of a peaceful future after decades of war. And yet, although it is not one of Central Asia's five republics, its future is inextricably linked with that of its neighbours. Many of the challenges that the United States and its coalition allies face in Afghanistan mirror those that confronted the USSR during its military occupation of 1979–89, when Stalin's successors waged war not only on nature, but also on anyone and anything they saw as a threat to Communist hegemony.

The Soviet Union eventually gave up its military efforts to sustain a Soviet-friendly but deeply unpopular Afghan government and left after ten years of conflict, heavy casualties, massive aid spending, shifting alliances and stalemate, on and off the battlefield. But before the Soviet army departed, the Kremlin's extensive planning and preparation for withdrawal had allowed the last and deeply unpopular Afghan Communist government to survive for a time. It fell in 1992, not long after the USSR collapsed. A bloody power struggle then followed among Afghanistan's many social, ethnic and religious tribes and factions, eventually paving the way for the Taleban takeover, al-Qaeda training camps and the September 11 attacks on the United States. These attacks prompted a renewed intervention by foreign powers, led by the United States, with the support of the British.

The US and its allies are fighting many of the same insurgents that America supported during the Soviet intervention in Afghanistan. Meanwhile, the coalition is trying to retain the uncertain allegiance of those who remain on the Western side, while checking the spread of Islamist terrorism from or within this region. The spectacular killing of Osama bin Laden in nearby Pakistan in 2011 revealed the depth of US intelligence and surveillance, while also clearly exposing the lack of trust between the Western allies and the Pakistani government.

The Soviet legacy still shapes not only Afghanistan's battlefields but also its daily life. Remnants of that era include mines, unexploded ordnance and other items useful to insurgents who make homemade bombs, known as improvised explosive devices (IEDs). The Soviet period bequeathed to Afghanistan large quantities of military hardware – some still functional – which attests to the Kremlin's abandoned efforts to strengthen Afghan security forces against insurgents. The United States and its allies now pursue that aim.

Afghanistan is shaped roughly like a clenched fist with the thumb extended out to the north-east. This 'thumbs-up' shape has never been indicative of good luck to anyone foolish enough to intervene in Afghan's troubled internal affairs. It is a rugged and mountainous country. The north-western, western, and southern border areas are mostly composed of desert plains and rocky ranges, whereas the south-east and north-east borders rise progressively higher into the major glacier-covered peaks of the Hindu Kush, an extension of the western Himalayas. Only the northern border is formed by a river, the Amu Darya. Afghanistan is linked with the Central Asian republics through trans-boundary water resources, almost 40 per cent of its territory being within the Aral Sea Basin, along with 33 per cent of its population.

Many of Afghanistan's major rivers are fed by mountain streams. The Amu Darya on the northern frontier has a number of significant tributaries that rise in the eastern Hindu Kush. It is the only navigable river in Afghanistan, though ferry boats can cross the deeper areas of other rivers. The Harirud River rises in central Afghanistan and flows to the west and north-west to form part of the border with Iran. The long Helmand River rises in the central Hindu Kush, crosses the south-west of the country and ends in Iran. It is used extensively for irrigation and agriculture, although in recent years its water has experienced a progressive build-up of mineral salts, which has decreased its usefulness. Most of the rivers end in inland seas, swamps or salt flats. The Kabul River is an exception. It flows east into Pakistan to join the Indus River, which empties into the Indian Ocean. Afghanistan has a few lakes and salt marshes, particularly in Helmand Province on the western border with Iran.

Prior to the civil war, less than 10 per cent of the country's hydroelectric potential had been developed. After the war began, hydroelectric production dropped off almost completely, as turbines were destroyed, floodgates blown open and transmission lines brought down. By the mid-1990s, private diesel-fired generators were about all that remained of 75 years of electricity development. This is why the supply of hydro-generated electricity from Rogun in Tajikistan may become an imperative towards helping the recovery of the post-conflict Afghan economy. But trans-boundary water issues must remain at the forefront of all future development plans.

Studies have revealed that until at least 2000BC, the land of Afghanistan was covered with cedar-rich forests and had a completely different environment and ecosystem from today. But the Afghan ecosystem has seen a marked deterioration over the past two decades, exacerbated by war.

Having said that, environmental issues in Afghanistan pre-date the political turmoil of the past few decades. Forests and wetlands have been depleted by centuries of grazing and farming, practices which have only increased with modern population growth. In Afghanistan, protection of the environment goes hand in hand with economic survival. With 80 per cent of the population dependent on herding or farming, conservation of the environment is critical to the economic well-being of the people.

The population depends on forests for fuel wood and the revenue generated from the export of pistachios and almonds, which grow in natural woodlands in the central and northern regions. Due to the war, the Badghis and Takhar provinces have lost more than 50 per cent of their pistachio woodland;

residents and military forces alike have used wood for fuel, and soldiers have cleared trees which could have provided hiding places for ambushes by the Taleban.

Denser forests in the eastern Nangarhar, Kunar and Nuristan provinces are also at risk from harvesting by timber barons. Although the logging is illegal, profits from exporting the timber abroad are very high. As forest cover decreases, the land becomes less productive, threatening desertification and the livelihood of the rural population. Loss of vegetation also creates a higher risk of floods, which not only endanger the people, but also cause soil erosion and decrease the amount of land available for agriculture. To make matters worse, with little government control to discourage hunting, and with the habitat disappearing because of conflict and drought, much of the country's wildlife is at risk.

Afghanistan has long been a land of marginal environmental conditions, too dry and too cold for much life to thrive productively. Thousands of years of environmental stress at the hands of the country's people have dramatically altered the landscape and caused extensive environmental destruction, which cannot be blamed solely on the ten years of Soviet intervention. Because the Afghan people lack the financial means to purchase fuel, they must cut trees, uproot shrubs and collect dung for burning. Domestic animals overgraze the plains. The result is extensive soil erosion by water and wind. Long-term irrigation without proper flushing has added salt to much of the arable land and destroyed its fertility. Polluted water supplies are common, except in the high mountain regions, where few people live permanently in any case.

The legacy of landmines in Afghanistan, created as a result of constant war, is the worst environmental nightmare of all. The presence of more than ten million landmines in the country makes it the world's most deadly minefield. The daily death toll due to these devices is about 20 to 30 people, mostly children and civilians.

Smog is also a problem in most of Afghanistan's urban areas. Even though the country doesn't have an industry which can create air pollutants, it is a victim of trans-boundary air pollution originating in the Aral Sea Basin, including from industrial zones in Iran, Turkmenistan and Uzbekistan.

Chemical weapons developed by the Soviets on Vozrozhdeniye Island and elsewhere were also used during the Afghan war with the Soviets, causing severe short-term damage to the environment and ecosystem. All in all, it can be concluded that the environment in Afghanistan is in deep crisis. The problems not only affect the people of Afghanistan and their ecosystem, but the whole world. Once any of the environmental components are lost, recovery is almost impossible.

Two-thirds of the landscape of Afghanistan is occupied by mountainous terrain with little or no vegetation, typical of an arid country. For this reason, the vegetation in these areas plays a vital role in the ecosystem, because half of the remaining parts of the country's landscape are deserts, which are hostile environments. The rest are farmlands and pastures. At present, only 6 per cent of the 15 per cent of agricultural land in Afghanistan is under cultivation. In the past 20 years, the agricultural areas have been drastically decreased. It is estimated that during the lengthy periods of conflict, Afghanistan has lost 30 per cent of its farmland and pasture, either by abandonment or degradation. This has led to a drastic form of climate change in the region. Compared to 1979, agricultural output has fallen by 50 per cent. To compensate for this loss, rural people started to exploit the free natural resources of their environment. The end result of this process has been a disaster for the few remaining natural forests, which were cut and smuggled to Pakistan. Deforestation, floods and avalanches have added to the devastation.

But in terms of regional stability, the main problem posed by Afghanistan relates to its use of water carried by the Amu Darya River, which is shared with Tajikistan, Uzbekistan and Turkmenistan, and eventually flows into the southern section, or Large Aral Sea. When, or if, the security situation stabilises in Afghanistan, much of the country's development will focus on irrigated agriculture, which will, in turn, mean increased use of the water resources available from the already overexploited Amu Darya.

The total population of Afghanistan now numbers around 35 million, out of which 8 million live in northern Afghanistan, equally split between the two great river basins – the Panj-Amu Basin and the Northern Basin. With 48 inhabitants per square kilometre, population density in irrigated areas situated along the main tributaries of the Amu Darya River and the river itself is high.

In recent years there have been a number of high-level talks about water use between Afghanistan and Tajikistan. These talks covered issues like integrated water resource management and planning in the Amu Darya Basin in general, and the installation of hydrological stations and bank protection measures along the Panj River in particular. In spite of the goodwill prevailing at these meetings in the past, little concrete action followed the good intentions.

For too long, the international community has chosen to ignore the difficulties encountered by regional governments in trying to protect and conserve upstream areas and their crucial, life-supporting function along trans-boundary rivers. In the case of the upper Amu Darya Basin, geopolitical issues have and will continue to dominate its use of natural resources. Caught in the middle of intensifying east-west trade, increased south-north drug trafficking and the first Taleban reported in 2009 to have crossed the Panj River into the Gorno-Badakhshan Autonomous Oblast, the role of some 8 million Afghans living in the north and north-east of the country – including some 300,000 Pamiri, who live between Tajikistan and Afghanistan – cannot only be limited to survival strategies that rely on their well-known resilience. These people need to be supported so that they can face future challenges and achieve a satisfactory living standard together with a positive outlook for future generations.

Climate change has already affected the availability of irrigation water in the Amu Darya Basin. Due to earlier snow melting, spring floods increase in size, and a shortage of water consequently occurs in summer and early autumn, increasing the risk of acute water shortages, particularly in years of drought.

The mountains of Afghanistan have always served as a natural reservoir and storage facility for water. More than 80 per cent of Afghanistan's water resources originate in the Hindu Kush Mountains. The snow accumulates in the winter and melts in the spring. This, along with the melting of the glaciers in the summer, feeds important rivers like the Amu Darya. The Amu Darya Basin alone holds more than 55 per cent of Afghanistan's water resources. Unfortunately, drought and the warming of air temperatures due to climate change have reduced the size of the glaciers in Afghanistan. Major glaciers in the Pamir and Hindu Kush have shrunk dramatically, while smaller ones have vanished completely.

Nationwide, the majority of Afghan households do not have access to safe drinking water. Water contamination due to unsafe sanitation is a major problem. Many water sources are contaminated with dangerous bacteria like E. coli, causing widespread death and illness, particularly among children and the elderly. It is common for household waste to be discharged directly into local streams and rivers. In addition, chemicals and other effluent pollutants are present at dangerous levels in most bodies of water. Even in the capital, Kabul, there are places where the water quality is so poor that it is dangerous to drink. A water law aimed at tackling some of these problems is still snarled in the legislative pipeline.

Speaking in Kabul in January 2011, Pekka Haavisto, the chairman of the UNEP Afghanistan Task Force, said: 'Water is a major problem in rural and urban areas due to water scarcity, mismanagement and damaged water systems. Although the country as a whole uses less than one-third of its potential 75,000 million cubic metres of water resources, regional differences in supply, inefficient use and wastage mean that a major part of the country experiences scarcity.'

Haavisto continued: 'Water quality, quantity and its guaranteed availability to all people regardless of income or social status is one of the most pressing challenges facing not only Afghanistan but also the world community today.'

Afghan government officials have also expressed concern. 'The water issue is becoming a serious problem, and the last four years of drought added to an already big issue,' Yusuf Nuristani, the Afghan minister of irrigation, water resources and environment, told a 'World Environment and Water Day' conference in Kabul in late 2010. He stated that only 20 per cent of Afghans nationwide had access to safe drinking water in both cities and rural areas.

The minister said water mismanagement was widely practised in the country and that, as a result of prolonged conflict, most water channels and other systems had suffered greatly. 'Restoration of water resources is one of the priorities of the government,' said Nuristani, noting that his ministry was now working out a strategy to bring about the improved management of water resources.

The special representative of the UN secretary general for Afghanistan, Lakhdar Brahimi, said the water issue was more than an environmental problem in the country. 'Water is, perhaps, the most precious resource in Afghanistan, and so it can be a source of conflict.' Brahimi said that much of the conflict in the country was the result of land disputes. 'Land rights do not mean much without water rights,' he said, stressing that one of the most important tasks facing the country was to impose order and the rule of law over land and water rights.

Samandar, a 40-year-old peasant from Andarab, a district of the northern province of Baghlan, agreed, saying he had lost a son and a brother to a water dispute in his village. 'They were killed by farmers from a nearby village,' the father of eight told the conference in Kabul. He said he believed that over 70 per cent of the tensions and anxieties affecting his village arose from disputes over the distribution and use of scarce water required for irrigation.

Following two decades of war, Afghanistan faces many major environmental problems, mainly caused by the degradation of water tables and the shrinking of wetlands and deforestation. According to the Afghan Ministry of Irrigation, Water Resources and Environment, some 40 per cent of forests have been cut down, while desertification and pollution of underground water represents another serious challenge.

But the road ahead is a long one. The ministry says it has

finalised a three-year development budget to cope with its most pressing problems. 'We have estimated close to $700 million for our three-year development plan,' the minister said, noting that $55 million of that sum had already been pledged by donors for the current year.

The increased use of water from the Amu Darya River may have serious consequences for inter-state relations with Central Asian neighbours. Increased agricultural production will obviously require water for irrigation, and some experts have told me that post-war Afghanistan could double the amount of water it currently uses. Any cropping that takes place in the northern part of Afghanistan will draw water from the Amu Darya, and such a situation will invariably create tension and enhance the risk of conflict with the Amu Darya's downstream users, who already encounter water availability problems. Consequently, the international community and development agencies must resist the temptation to assume that Afghanistan's development needs automatically outweigh those of the Central Asian states.

Opium has flourished in Afghanistan since the time of Alexander the Great, when it was used as a medicine. But under the Taleban production increased spectacularly, to the point where Afghanistan supplied 80 per cent of Europe's heroin. After several years of unofficial tolerance and profit from the crop, the Taleban virtually eradicated Afghanistan's opium crop in 2000, following an edict by Mullah Mohammad Omar, the Taleban leader. The Taleban ban on poppy growing slashed Afghan opium production by 94 per cent that year. But the ongoing war and the need for cash has led to a change of heart. Desperate for funds to finance their vicious guerrilla campaign, the Taleban have permitted a resurgence of the poppy crop and are masterminding the supply of heroin to the international trade.

Out of Afghanistan's 29 provinces, ten now grow poppies. Of these, the richest are Helmand in the south, where much of the current fighting is concentrated, and Nangrahar in the east. In a recent speech, the agriculture minister in Afghanistan told Afghan farmers that they should switch from cultivating poppies for the heroin trade to pomegranates. He said that there was growing demand for pomegranate juice in the West because of its anti-oxidant qualities, and Afghan soil is ideal for pomegranates. At face value this is a sound idea, and one that the Western powers are likely to seize upon to boost the Afghan economy, disrupt the poppy harvest and thereby sever the links to the international drugs trade from which the Taleban earns most of its money. The problem with this idea is of course water, or the lack of it. Pomegranates are a thirsty crop, and a huge increase in water use will have inevitable and potentially explosive consequences downstream. It would be a great pity if we managed to secure an end to conflict in Afghanistan only to see two or three mini Afghanistan-type conflicts break out downstream because of acute water shortages.

Water is also an issue for other popular crops. Afghanis love to eat melons for their sweet taste and texture. However, their popularity stems from another, less conventional use. In northern Afghanistan melons are used to prolong the high of cannabis users. Afghanis believe that when eaten while smoking hashish, they get a longer high and twice the pleasure in the summer when the melons are particularly ripe. They claim that melons are not only good for your health, but they also keep the drugs from having a bad effect.

In northern Afghanistan, many farmers grow melons and watermelons as a cash crop. Different varieties include makayee, loblayee and gorgak. A gorgak melon is very small and green, but it is the sweetest of them all. Local farmers only grow this variety for their own use and never sell them, as to do so is considered unlucky. Farmers can expect to harvest 400 to 600 melons or watermelons per jerib, which is the way Afghanis measure land. One jerib is equal to around two hectares. Melons can sell for 30 to 70 Afghani each, giving an income of approximately $500 US per jerib. No exact statistics are held, but surveys by the agriculture department in the Balkh province indicate that more than 100,000 jeribs of land across northern Afghanistan are planted with melons. It is safe to assume that this area will increase significantly in post-conflict Afghanistan, increasing the use of water exponentially.

CHAPTER TWELVE

Turkmenistan and the Golden Age Lake

Far from learning the lessons of the Soviet era's eco catastrophes, some of the nations of Central Asia seem determined
to repeat the mistakes of their forbears. Nowhere is this more
apparent than in the downstream nation of Turkmenistan,
where a colossal artificial lake is being constructed.

The 'Golden Age Lake', or 'Lake Turkmen', in the Karakum
(black sand) desert will, when completed, be 130 metres deep
with a capacity of 123 cubic kilometres of water. It is located in
a salt depression in the north of the country called Karashor
which was formerly a river bed of the Uzboy, a tributary of the
Amu Darya. It will have a surface area of 2,000 square kilometres which, to put it in context, will make it 35 times larger
than Loch Ness. It is currently the world's biggest engineering
project, costing an estimated £12 billion.

To achieve this feat of construction, Turkmen engineers have
built a 30m-high and 2km-wide dam across a natural depression at Karashor to collect water and form this inland sea. The
Turkmen government claims that there are two aims of the
Golden Age Lake project: first, to collect excess water run-off
from the irrigation of cotton fields, which will be cleaned and
used to irrigate nearly 400,000 hectares of new pastures and
orchards; and secondly, to create and maintain a lake which
can then be used for commercial fishing and tourism, as well as
a source of water for irrigation. They say that when the project
is fully completed and operational the lake will quadruple the
current area of irrigated agricultural land in Turkmenistan.

Inevitably, the construction of the lake is a contentious issue
with Turkmenistan's neighbours because it will divert water,

once again, from the Amu Darya. Even before the lake has been fully completed, the levels in the lower reaches of the Amu Darya have dropped noticeably because of the project. Probably as a direct consequence of this in 2001, a large number of people in both Karakalpakstan and Khorezm were forced to flee their homes due to a sudden acute shortage of both irrigation and drinking water. Thousands migrated to the neighbouring regions of Turkmenistan and Kazakhstan.

Circumstances are made worse by the fact that water issues lie at the top of Turkmenistan's political agenda, which it tends to view solely as a domestic question. 'Water is our affair and no one else's,' seems to be the prevailing attitude. As a result, Turkmenistan refuses to be involved in any regional meetings with its Central Asian neighbours on management of water resources, apart from IFAS, the International Fund for Saving the Aral Sea, where it sniffs the prospect of significant funding from the global donor community.

Meanwhile, the tightly controlled Turkmen press, which slavishly follows the government line, continues to ignore the geopolitical implications of the Lake Turkmen project and instead churns out Soviet-style propaganda, claiming that the lake will create vast green pastures and orchards as well as creating untold wealth as a regional holiday resort. President Kurbanguly Berdymukhamedov claims the Golden Age Lake plan shows his country is 'preserving nature and improving the environment'.

Unsurprisingly, inter-state relations have been affected by the construction of the lake, particularly between Turkmenistan and Uzbekistan. Relations are historically tense between these two states because of the shared resources of the Amu Darya Basin, mainly because the Amu Darya River is the primary source for the waters that irrigate the Uzbek agricultural sector, including the giant cotton industry upon which it heavily relies for exports and foreign currency. If the completed Golden Age Lake project further reduces the water available to Uzbekistan, it will most certainly put even more strain on relations and will affect stability in Central Asia.

Worse still, some experts argue that due to Turkmenistan's

poor soils and arid climate, the result of the project will be a massive dead lake that will exacerbate desertification and salinisation. Experts fear that this large-scale project will wreck the already fragile ecosystem and that water will simply evaporate in the desert. They say that lessons have not been learned from the irrigation channels and canals built during Soviet times, with their unlined earth walls, and that these mistakes are being repeated. They argue that unprotected earth banks in permeable soils will allow the mass seepage of water polluted with fertilisers and insecticides and the consequent salinisation of big areas or farmland. In other words, almost an 'Aral Sea Mark II' or a new 'Dead Sea'.

The project was begun by Turkmenistan's former president, Saparmurat Niyazov, who was renowned for his extravagant and huge building projects. Niyazov died in 2006. Mr Berdymukhamedov came to power vowing to break with the past. But he has already approved several $1 billion projects for the capital, Ashkhabad, including a new five-star hotel, government buildings, a new stadium and a 'Palace of Happiness' for weddings. He is gradually trying to replace the personality cult that surrounded his predecessor with one of his own.

In July 2009, President Berdymukhamedov, wielding a spade, opened up the first tributary to bring water to the Karashor natural depression in the Karakum Desert, the nascent Golden Age Lake. He told the crowd that the lake would make the desert bloom. 'Our initiatives to provide water and environmental security demonstrate that Turkmenistan is making huge efforts to contribute to common work on preserving nature and improving the environment,' he said. The water from the canals, he added, would attract wildlife and open up new land for agriculture. Village elders in traditional clothing helped the water flow into the new channel. After the opening ceremony, Mr Berdymukhamedov mounted a richly bejewelled white horse to ride back to the helicopter which brought him from his palace in Ashkhabad. The Turkmen government, on its website, boasted that the project 'would go down in the history of the epoch of New Revival as one of its brightest pages'.

Work on the project began in 2000, with the construction of two canals which bisect the country. Thousands of smaller feeder channels were built to funnel water from Turkmenistan's irrigated cotton fields to the new lake. Treatment plants are planned to clean the water, although it is far from certain that this will be successful.

I visited Turkmenistan in April 2010. The capital, Ashkhabad, is a show city, rather like its closest similar cousins in Dubai or the Emirates. Marble palaces have sprung up in vast numbers, but they line wide boulevards and have great gaps between each building, giving a feeling of spaciousness. All conform to a similar pattern of design, apparently masterminded by a French architect, with Greco-Roman pillars, great statues and fountains everywhere. The overall effect is like walking into ancient Rome at the time of Emperor Hadrian, with all its sumptuous pomp and importance.

It is only gradually that the first-time visitor to this Central Asian capital realises that there is something wrong – something missing. There are very few people in evidence. The wide streets are empty, save for an army of women street-sweepers. There is at least one of these positioned every 100 metres, sweeping purposefully at the dust or puddles with birch brooms. A male supervisor every 200 metres keeps an eye on them. This activity can be seen night and day, and is clearly a major source of work in a city with over 40 per cent unemployment.

But the street sweepers, the ubiquitous MNB (the KGB's successor organisation) and the military appear to be the only residents of this new marble paradise. Ordinary Turkmens are kept at bay by the police. Endless roadblocks ensure that mere citizens cannot penetrate the marble show city, where the multi-million-dollar buildings lie locked and empty, simply there to impress visiting dignitaries.

Turkmenistan has the smallest population of the five former Soviet republics in Central Asia. The government is seen as the region's most autocratic, but the strict isolation imposed by eccentric dictator Saparmurat Niyazov has lifted somewhat after his death. It is still effectively a one-party state dominated by the Democratic Party of Turkmenistan, led by Niyazov until his death in December 2006. The late leader styled himself Turkmenbashi, or Father of the Turkmen, and made himself the centre of an omnipresent cult of personality. Mr Niyazov, who was made president for life in 1999, spent large sums of public money on numerous grandiose projects but not on social welfare.

His influence spread into every conceivable area of life in the republic. Turkmens were even expected to take spiritual guidance from his book, *Ruhnama*, a collection of thoughts on Turkmen culture and history. Turkmenistan is the most ethnically homogeneous of the Central Asian republics, the vast majority of its population consisting of Turkmens. There are also Uzbeks, Russians and smaller minorities of Kazakhs, Tatars, Ukrainians, Azerbaijanis and Armenians. In contrast to other former Soviet republics, it has been largely free of inter-ethnic hostilities. However, strong tribal allegiances can be a source of tension.

With foreign investors keeping away, the Turkmen economy remains underdeveloped. The country has been unable to benefit fully from its gas and oil deposits because of an absence of export routes and a dispute between the Caspian Sea littoral states over the legal status of offshore oil wells. Turkmenistan produces roughly 70 billion cubic metres of natural gas each year, and about two-thirds of its exports go to Russia's Gazprom. A protracted dispute between the two countries over the price of gas ended in September 2006, when Gazprom agreed to pay 54 per cent more. Turkmenistan has since made efforts to break out of Russia's hold on its exports. It has opened major gas pipelines to China and Iran, and is considering taking part in the Nabucco pipeline – an EU-backed project designed to provide an alternative to Russian gas supplies to Europe.

Mr Berdymukhamedov became acting president after authoritarian leader Saparmurat Niyazov died in December 2006. His nomination for the presidency surprised observers because under the constitution the post should have gone to People's Council chairman Ovezgeldy Atayev. However, after

Mr Niyazov died, Mr Atayev became the subject of a criminal investigation and was sacked.

Berdymukhamedov was sworn in as president in 2007, with 89 per cent of the vote. (There were six candidates, all from the Democratic Party of Turkmenistan, as exiled figures from the opposition were banned.) Human rights groups and Western diplomats condemned the election as rigged. Weeks later, the president was chosen as chairman of the People's Council, Turkmenistan's highest legislative body. He was the only candidate.

Since the death of Turkmenbashi, his successor and former dentist, Korbanguly Berdymukhamhedov, has ordered the display of massive photos of himself on the outside and inside walls of every public building. It is rumoured that even the infamous Independence Arch, which features a pure gold statue of Turkmenbashi which swivels to face the sun, is due to be dismantled soon. The Turkmens laughingly refer to it as the Turkmen Grill because the statue gently turns in the heat, and clearly believe the former president was bonkers. They are careful not to express any such doubts about the current incumbent, however.

There is a sort of informal curfew in Ashkhabadat 11 p.m. every evening. Although not promulgated by any law, citizens out on the streets after this hour are likely to be detained by the police or MNB and questioned closely as to why they are not sitting at home watching one or other of the wholly government-controlled TV channels which show endless news reports of the president and his important goings-on.

Turkmenistan earns an estimated $10 billion a year from Russia's Gazprom for the sale of gas, so the president had loads of cash to spend on his pet projects, such as the new marble city of Ashkhabad. But in this quasi-Communist regime where everything is centrally controlled by the president himself, even electricity, gas and water are supplied free. Apartments are allocated on a rent-free basis. The first 25 litres of petrol a citizen uses every month is free; after that it is charged at only 20 US cents per litre. But it is mostly the privileged elite who benefit from these policies. Most of the population live at a level of poverty and with a life expectancy similar to some of the poorer nations of central Africa.

Tourists and visitors are mercilessly ripped off. Tariff charts in the national museum, for example, show that locals pay 10 cents while tourists have to pay $10. Government-owned restaurants have 20-page menus of low-quality food, badly cooked and served by sullen staff, but at prices that would make Albert Roux choke on his foie gras. English translations are available in some, but these tend to provide the only entertainment of the evening, describing tender veal, for example, as 'sentimental calf'. A glass of low-grade Moldavian wine can cost over $50.

The national museum now has an entire floor devoted to President Korbanguly Berdymukhamhedov, who is apparently a great cook, an acclaimed writer, an Olympic horse rider, accomplished rock guitarist and much else besides, according to the exhibits, which consist mostly of rather boring and ridiculously posed photos. All hotel rooms in Ashkhabad are bugged. The only five-star accommodation in the city – the President Hotel – has blocked computer access in every room, forcing guests to use the business lounge, where only two computers are available, only one of which has internet access, ensuring that all outgoing and incoming email traffic can be monitored. The President Hotel allocates its rooms on a language basis, so that English-speakers are always put on the 10th floor and Kazakhs on the 11th floor, to make bugging and monitoring easier. All mobile calls in Turkmenistan are monitored.

The people are oppressed to a degree unseen outside North Korea. By law, all women, and even young girls, have to wear long skirts to avoid showing their legs and causing offence. A relentless crackdown on the state-owned media ensures that nothing critical or even faintly disturbing can be reported. There is little or no crime, because of the constant presence of MNB agents, soldiers and police, but any crime, or even any motor accident, will never be reported by the media, who are fed an endless diet of how wonderful Turkmenistan is and how lucky its five million inhabitants are to live there. MNB informants infiltrate all levels of society. Those who seek to dissent are punished by torture, imprisonment, house arrest, surveillance and incarceration in psychiatric facilities. Although things have relaxed slightly since the death of Turkmenbashi, there is still a pervasive feeling of fear and a marked reluctance on the part of the average citizen to talk to foreigners.

But in the bizarre new city that Turkmenbashi created in Ashkhabad, a fabulous new Russian school was opened by President Medvedev and President Korbanguly Berdymukhamhedov. It cost a staggering $22 million, all paid by Gazprom, to a specification drawn up by Berdymukhamhedov. After a great showpiece opening, it was quickly locked up again and remains empty to this day. Locals told me that the president had even designed the classroom litter bins, which were specially manufactured and cost over $500 each.

There is no private enterprise, and there are no private businesses. There are no taxis; Berdymukhamhedov recently purchased a fleet on 1,200 taxis, but they remain locked up in garages around Ashkhabad, unused. Ordinary citizens hitch lifts with car owners, Moscow-style, and pay them a few cents for each journey. However, a new fleet of Boeing 777s purchased by Berdymukhamhedov ply regular direct routes to Frankfurt, London, Delhi and Almaty in Kazakhstan.

Despite the lavish marble buildings and multicoloured fountains and statues that decorate every corner and roundabout, the city lacks soul. The only part of Ashkhabad that has a feeling of reality about it is the old Soviet part, which buzzes with life, people, cars and activity. It is here where the majority of the population live. A Lithuanian diplomat friend who was staying in the President Hotel asked me if I wanted to accompany him on a visit to the old city, where his cousin lived. I readily agreed. His cousin came to collect us at the hotel in his rather dilapidated Toyota and told us he'd been stopped four times at various roadblocks before finally getting through. We were stopped again twice on the way back out of the new city.

Finally, we got past the last of the KGB roadblocks. We left the tree-lined marble boulevards of new Ashkhabad and entered the more familiar grey and concrete world of the former USSR. It was reassuring to see traffic and people again in the old Soviet parts of Ashkhabad. We wound our way through fairly busy streets until we arrived at some decrepit housing blocks on the outskirts of the city. We were being taken by the diplomat's cousin to visit a man who bred rare Japanese Akita Inu dogs. As a lover of dogs, I was intrigued that such an enormous and rare canine breed, which are extremely expensive in the West, could be reared in Ashkhabad.

We drove into a muddy square surrounded by corrugated iron sheds. Shoeless children were playing with mangy-looking dogs. Elderly women were cooking on small coal fires set on the ground outside their front doors. A single cold-water tap in the middle of the square was being used by women from various dwellings to wash plates and fill kettles and pans. Clearly, this was the only source of water in the block.

A group of men greeted us with wide smiles and ushered us towards the back of the buildings, where loud growls and barks could be heard. We passed a single camel, its front feet hobbled with rope, munching contentedly on its cud. From here, on this elevated ground on the city outskirts, there was a clear, panoramic view of Ashkhabad. It was strange to be standing in the midst of such abject poverty, where a single tap provided the only source of water for several families, with holes dug in the ground for toilets, and yet we could see spread out beneath us the lavish, marble fountains, gold onion-domed mosques, flowerbeds and skyscrapers of the new city that had sprung up in the past two decades, a virtual no-go area for all but the elite.

One of the men slid a bolt from its moorings on a rickety, corrugated-iron gate and ushered us into a narrow, dank passage formed by a gable wall on one side and a row of large cages on the other. The pungent smell from the kennels made my eyes water. Inside the cages, huge white dogs snarled and snapped as we approached. It quickly became evident why my friend's cousin was involved in breeding Akita Inu dogs in Ashkhabad. These were fighting dogs. Their ears and tails had been roughly docked at birth, and each of the adults bore livid scars on their muzzles and flanks. Dog fighting is legal in Turkmenistan and is a major spectator sport, involving large amounts of money in wagers. It is a savage spectacle, and dogs are routinely torn to shreds in the weekly tournaments. These animals are bred to be killers. They are fierce and fearless.

One of the bigger dogs was particularly displeased to see strangers walking close to his cage. He charged headlong into the iron bars, uttering a blood-curdling roar of rage. The cage shook alarmingly and I turned on my heels, ready to make a run for it, which made the local Turkmens laugh and slap their thighs in glee. A young boy shouted at the big white dog in the next cage and slowly unlocked the padlock that secured the kennel. He reached in and grabbed the growling beast's collar, tying a stout rope around it and pulling it out into the passageway. It stared at me and snarled, and got a sharp smack on the nose from the young man for its trouble. Another teenager extracted an even bigger dog from a neighbouring cage, and soon we set off out of the kennels with several fighting dogs in tow.

Behind the kennels was a steep, grassy embankment. This was obviously the exercise area for the dogs. The boys started to run quickly up and down a pathway, the huge Akita Inus loping along gamely at their sides. Soon, we were joined by two of the smaller pet dogs from the housing scheme, who ran alongside their huge cousins, barking excitedly. One of the boys proffered his rope to me, indicating that I was to run the big Akita along the path. I held my hand down for the dog to sniff, hoping it would not disappear in a single bite, but the heavily scarred Akita seemed unfazed and wagged the stump of his tail happily. I jogged off, the massive hound almost pulling me along, its pink tongue lolling from its jaws and flecks of frothy spit spattering my clothes. The local men cheered and applauded in the background.

Later, covered in white dog hairs and smelling faintly of Akita, we were taken to a restaurant which the diplomat's cousin said was one of the few examples of private enterprise in

Ashkhabad. It was in the heart of the old city, next to a major road and, judging by the busy car park, was a favourite stop for lorry drivers. Our friend explained that, unusually, the government, who own everything in Turkmenistan, had agreed to lease the premises for use as a restaurant. The narrow room was filled to overflowing already, but extra chairs and a table were quickly found for us. Soon women from the kitchen were carrying baskets of freshly baked bread and great heaped platefuls of fried fish, all washed down with endless toasts of chilled vodka.

On our way back to the President Hotel, we were stopped at the perimeter of the new city by a KGB roadblock. They instructed us to turn around and said that no one was permitted to enter the zone because UN Secretary General Ban Ki Moon was staying at the President Hotel. I told my interpreter to explain that we were also staying in the President Hotel, and indeed we were due to join Ban Ki Moon on his visit to the Polygon in Kazakhstan later that week. The senior KGB officer looked at our scruffy old Toyota and our local driver and shook his head in disbelief, repeating his shouted threat for us to do a U-turn and stop trying to enter the new city. Our driver said he knew another way to get to the President Hotel and we began a long circuit of the city. Eventually, around one in the morning, we again approached the perimeter of the security zone and again were confronted by a KGB roadblock determined not to let us pass.

This was the last straw. I was in Turkmenistan on official business, after all, and was fed up being harassed by these pumped-up bureaucrats. I jumped out of the car and asked Anna, my interpreter, to join me. Marching up to the nearest KGB official, I pulled out my diplomatic passport, opened it at the page written in Russian and waved it under his nose, shouting that I was a guest of his president and he had damned well better let me get back to my hotel or suffer the consequences. This seemed to have some sort of impact because several officials now swarmed around us, leafing through my diplomatic passport while making anguished calls on their mobile phones.

Sure enough, within minutes a black limousine pulled up, disgorging an official from the Ministry of Foreign Affairs who confirmed our identities and told me, my interpreter and Vytautis, our Lithuanian diplomat friend, to get into his car. Our friend in the Toyota was ordered to make himself scarce, but not before I noticed one of the KGB officials writing down his car registration number in a small notebook. No doubt he would be questioned about his night out with us later.

As we drove through the deserted streets of the new city in the ministry limo, the government official who had come to our rescue started gently probing us for details of where we had been and whom we had spoken to during the course of our visit earlier in the evening. It quickly became apparent that our disappearance from the President Hotel and the fact that the KGB had lost track of us had not gone down well, and a full-scale inquiry was now underway to discover what we had been up to. We, of course, played the role of dumb foreigners, saying we couldn't remember the name of the restaurant or even its location. We didn't mention the visit to the Akita Inu dog breeder.

This, however, was not to be our only encounter with the security services in Ashkhabad. The next morning, having a couple of hours to spare before our first official meeting, myself, Anna and Vytautis decided we would like to visit the famous Independence Arch, with its legendary pure gold statue of the former President Turkmenbashi. We had passed it several times on our various journeys through the centre of Ashkhabad and had noticed that there was a viewing platform near the top of the arch from which we hoped we could get a panoramic perspective of the whole city.

To ensure that we did not attempt any more 'disappearing acts' in Ashkhabad, the Ministry of Foreign Affairs had allocated us with a car and driver at our permanent disposal. Our thick-set and rather surly driver was waiting patiently outside the President Hotel, smoking a cigarette, one arm dangling nonchalantly from the window of the white limo. We clambered on board and instructed him to take us to the Independence Arch. Once again the streets in the new city were eerily deserted. We seemed to be the only car on the move. We passed rows of gigantic marble-clad buildings with huge show windows and incongruous signs in Russian proclaiming 'Baker', 'Butcher' or 'Hairdresser', although it was clear that each of these lavish premises was empty and locked. The cherry trees were a blaze of pink and white blossoms, and the air was rich with the perfume from endless beds of flowers.

Soon we arrived at the foot of the Independence Arch. It is called an arch, although its design more resembles a comic-book spaceship, with a great tower held aloft on three huge legs which straddle the main thoroughfare. There is a circular viewing platform, enclosed in glass, immediately above the arch, and another smaller platform with an outside balcony right at the top of the tower. On the very pinnacle of this spectacular construction is the golden statue of former President Turkmenbashi – Father of the Turkmens – a golden cloak billowing from his shoulders like the wings of an angel, his hands outspread to greet the sun.

It was President Niyazov – alias Turkmenbashi – who abandoned the traditional calendar and substituted months and days named after members of his family. A colossal sculpture of a book he wrote, entitled *Ruhnama*, adorns one of the main squares in Ashkhabad, its ten-metre-high pages slowly opening to reveal the Cyrillic script, while a disembodied voice booms out extracts from the learned text. The *Ruhnama* is a religious book, which the Turkmenbashi regime insisted must form the basis of the educational system in Turkmenistan and be given equal status with the Koran. Indeed mosques in Turkmenistan were required to display the two books side by side. The book was heavily promoted as part of the former president's personality cult – knowledge of the *Ruhnama* was even required for obtaining a driver's licence.

We made our way up the steps to what appeared to be a ticket office at the foot of the Independence Arch. A cigarette lay smouldering in an ashtray, but there was no one in sight. A military guard with a sub-machine gun pointed to the circular viewing platform high above us and indicated that we must take the lift up there to get tickets. We squeezed into a claustrophobically small lift with posters of the president peeling off its walls and trundled up to the first platform. Here, a surprised-looking official took our money and ushered us into another impossibly small lift to make the ascent to the top platform. He made it clear that he would remain behind, and we creaked and groaned our way up to the top of the tower, with occasional glimpses of the city receding far below us.

Emerging once more from the lift, we found ourselves on a circular, glass-encased platform with doors here and there allowing access to the outside balcony. Everything was in a state of serious decay. The carpets were damp and threadbare. The walls were rusty and wet. The windows were filthy. We were disappointed to discover that the doors to the balcony were locked and it was almost impossible to see anything of the city through the grimy windows. But as I wandered around the circular platform I suddenly uttered a shout of glee. One of the doors to the balcony was slightly ajar. We all clambered out onto the balcony, cameras at the ready.

As we surveyed the neatly laid-out streets and tree-lined boulevards far below, I began to notice a strange phenomenon. Immediately beneath the tower, no cars or pedestrians could be seen. And yet a few hundred metres away in an almost perfect circle around our tower, we could see police and KGB officials blocking the roads and pavements and holding back crowds of pedestrians and vehicles. We seemed to be in the middle of a total isolation zone. As I took photographs of this bizarre scene, I suddenly became conscious of a tiny ant-like figure far below us shouting and gesticulating. I called the others to come and look and we quickly became aware that he was shouting and pointing at us. By now, police, soldiers and KGB officials were running towards the base of the tower from all directions, all staring up at us and many waving guns and joining in the general clamour.

I decided that perhaps it would be wise if we retreated inside the building and made our way back down to ground level. As we descended in the lift, we could see a sizable posse was now sprinting towards the tower, shouting and pointing at us. It was all a little unnerving. I was beginning to think that some idiot with a gun might open fire on us at any moment, deeming us guilty of some obscure breach of security. In the event, as the lift came to a halt on the lower platform and we emerged, the posse suddenly stopped in their tracks, having spotted that we were foreigners. They turned their attention and wrath instead to the poor ticket clerk who had allowed us to ascend the tower, shouting at him to come down immediately to ground level. We all squeezed into the lift together, the ticket clerk visibly shaking.

As we emerged at ground level, the posse of KGB and police rushed forward and grabbed the ticket clerk, jostling and shouting in his face. Just then we noticed a cavalcade of limousines rolling past the Independence Arch. In the lead was the large black and white limousine bearing the UN flag. It was Secretary General Ban Ki Moon on his way to the airport. Obviously this was why security had been tightened, and cars and pedestrians had been held back to enable the cavalcade to pass unharmed. Clearly our presence on the balcony at the top of the tower had been a major security breach, with the KGB presumably thinking we were potential assassins! We really were lucky not to have been shot. I hate to think what happened to the poor ticket clerk. Despite our protestations and explanations, he was led away. High above us, the golden statue of Turkmenbashi slowly pivoted to greet the sun.

Back safely in our government limo, we instructed our driver to take us to the Ministry of Foreign Affairs, where I was due to meet the Deputy Minister Mr W. Hajiev. A small, podgy man with dyed black hair, he explained: 'We held a recent conference on the Golden Century Lake. A lot of subjective and superficial comments have been made about this lake. Turkmenistan has proposed various solutions such as supplying energy to upstream countries, in other words replicating the old Soviet scheme under new conditions.' We were sitting at opposite sides of a large table and the minister was surrounded by aides who noted down every word each of us uttered. A cameraman had positioned himself behind his tripod in the corner of the minister's office and was filming the meeting.

I said that Uzbekistan in particular was alarmed about the Golden Century Lake project, worried that it would further diminish their supply of water from the Amu Darya, but Mr Hajiev replied: 'In Turkmenistan we are working on the development of new irrigation systems which involve the careful husbanding of water. This is a long-term project to increase irrigation and yet reduce water use. There is a very low utilisation of water from the Amu Darya for households and industry. Most is used for irrigation. Turkmenistan has the necessary volume of water that it needs for irrigation. Our plans are to leave high amounts of water in the Amu Darya. We have conducted extensive research on pollution and groundwater use. We are constructing a system of channels to collect excessive water run-off following irrigation and direct it to cleaning chambers before pouring into Lake Turkmen. Your own European Commission officials have studied and approved this project,' he assured me.

Mr Hajiev continued: 'As far as Afghanistan is concerned, it is in the interests of the entire region to resolve the conflict there. Dams are currently being built by Indian companies in Afghanistan. We are sending energy to the Afghans. They are our neighbours, and we need to help. We have already helped to reconstruct railroads serving Afghanistan, and we do not create problems at border crossings to enable building materials to enter the country. We will continue to help in Afghanistan. The Amu Darya is flooding in parts, and no work is being done in that country to resolve this problem. We are supplying resources to help them, such as the transportation of humanitarian aid. In the Northern Provinces we provide extensive aid. We share an 806km border with Afghanistan. There is also a bridge being built over the Amu Darya to create a link to Uzbekistan, together with new road networks to the Caspian.'

Later that evening as we drove to the airport for our flight back to Frankfurt, I was startled to see my own face staring down at me from a vast TV picture projected onto the side of a skyscraper. Apparently every visiting politician who meets a

Turkmen minister is filmed, and the film is then projected on a constant loop onto the sides of public buildings for the greater edification of the Turkmen population.

It was an appropriately bizarre end to a bizarre visit, and an excellent way to prepare for the next leg of my travels to the Ili-Balkash Basin.

CHAPTER THIRTEEN

Ili-Balkhash Basin

Having failed to learn any lessons from the desiccation of the Aral Sea, history is now being relentlessly repeated in Kazakhstan. Ill-conceived hydroelectric power (HEP) schemes constructed during Soviet times in the Ili-Balkash Basin began a spiral of decline that has now been accelerated by HEP industrial and agricultural projects, mainly in China.

Lake Balkhash and its basin are situated in the south-eastern part of Kazakhstan, and the main head waters come from the north-western part of China. A small part of the head waters pass through the territory of Kyrgyzstan. The lake is the third largest body of freshwater in the world, covering an area of 413,000 square kilometres. The Ili River, originating in China, is of key significance for Lake Balkhash because not only is it a primary source of water for the lake, but it also preserves its rich biodiversity, while preventing land degradation and deforestation of the basin area. Moreover, water from the Ili River is used for producing hydro power, and in industries such as mining and agriculture. It is one of the largest ecosystems on earth. A fifth of the population of Kazakhstan lives in the Ili-Balkhash Basin area. Despite its rich natural resources, the area of the basin is the one with the lowest income rate and with the highest unemployment, poverty and sickness rates in the country.

The environmental problems in the area have now reached national and international proportions, and will soon rank alongside the Aral Sea in terms of their catastrophic impact if imminent action is not taken. The main environmental problem involves the pollution of the lake with heavy metals, mainly coming from a copper plant in Balkhash. It is reckoned that the lake is loaded annually with 25,000 tonnes of heavy metals from this plant.

However, as in the case of the Aral Sea, excessive abstraction of water from the lake and feeding rivers for irrigation and industrial purposes is causing drastic shrinkage and is having damaging consequences for the local environment. The Chinese, for example, use substantial amounts of water for irrigation, and this is slowly drying up the lake. The way water is taken from the basin is both uncontrolled and uneconomic, and is leading to a significant loss of biodiversity, land degradation and a loss of water. The first signs of desertification are emerging as well.

At present, the Chinese government is completing a programme of socio-economic development in its western autonomous districts. This includes the widespread deployment of water resources affecting cross-border rivers like the Irtysh and the Ili, aimed at the development of irrigated land cultivation, stock breeding and hydropower, and the provision of water to the oil industry and the rapidly growing population of Xinjiang, in Western China.

Enormous projects have been started involving the building of canals, reservoirs, hydroelectric stations and other hydropower constructions on these river arteries. The central government in Beijing is spending over $243 million annually on their construction. Inevitably, this has caused major problems with neighbouring countries in the region, as, for example, the Irtysh river basin includes the territories of China, Kazakhstan and Russia.

Kazakhstan has come off worst in this scenario. The construction of reservoirs and drainage canals on tributaries of the Irtysh in China will lead to a situation where the total water taken from the river will reach 3 billion cubic metres or 35 to 37 per cent of the total annual flow of that river system every year. This will cause serious harm to flora and fauna, and will also lead to a change of climatic conditions not only in Kazakhstan and Russia but also in Xinjiang itself. Desertification could be a direct consequence of China's actions.

Similar problems have arisen in connection with the Ili River and its tributaries, where again the Chinese government has commenced major construction work on around 90 different projects involving dams, hydropower and irrigation canals. According to various expert surveys, if the long-term programme of developing water resources in the Ili basin and its tributaries is implemented in China, the flow of the river in Kazakhstan will decrease to 40 per cent by the year 2050.

It has to be borne in mind that the Ili-Balkhash Basin is one of the most unstable in Kazakhstan in terms of both its water supply and its ecology. The quality of the water in the Ili basin is one of its most basic problems. There is heavy chemical pollution coming from China and also from run-off from nearby agricultural land. Recent research conducted in the Ili River in the Almaty region of Kazakhstan showed that there was a strong presence of heavy metals and oil products all coming from China. For example, it was found that in fishing reservoirs fed by the Ili, copper was present at more than 35 times the permissible level, iron was 6.3 times over, manganese was 6 times over and nitrites were 14 times over. This pollution led directly to the deterioration of water in Lake Balkhash, the disturbance of the ecosystem of the Balkhash River and the gradual drying up of lakes and water marsh areas in the Ili River delta. We are now witnessing an unfolding disaster similar to that which affected the Aral Sea, causing massive desertification.

So, according to the evaluation of experts, the continued excessive use of the Irtysh River by China will lead to an ecological disaster in eastern and southern Kazakhstan. Lake Balkhash, situated at the end of the Ili River, already has an acute shortage of water. The long-term consequences of this continued over-utilisation of water resources by China will be climate change, damage to the fishing industry, reduction of the yield capacity of agricultural plants, the consequent degradation of pastures and an increased concentration of toxic substances in local water supplies, rendering them useless for household use.

The Ili River is approximately 1,400km long and provides

80 per cent of the water flowing into Lake Balkhash after passing through the Saryesik-Atyran Desert. Flowing into Lake Balkhash, it forms a delta of 8,000 square kilometres, which supports an abundance of flora and fauna. The Kapchagai Reservoir, built along the middle reaches of the Ili River in 1966 and used for water storage since 1970, allowed the development of irrigated agriculture along the lower reaches of the river. This reservoir serves for HEP generation and for irrigation water supply. Since its inauguration, water use has increased along the lower reaches of the Ili River, and the subsequent decrease in the river's inflow deeply affected the lake's environment and water system.

Lake Balkhash

Lake Balkhash is a terminal lake that lies in Eastern Kazakhstan. The lake is the largest moderately saline lake of Central Asia and has a surface area of over 16,000 square kilometres, a length of 600km and its width varies from 5km to 70km. Its drainage basin of approximately 413,000 square kilometres is situated in south-eastern Kazakhstan (85 per cent) and north-western China (15 per cent). The average depth of the lake is 6m, though its maximum depth reaches 26m, and it is normally frozen between November and March. The Ili-Balkhash Basin is home to over 3.2 million people, many of whom live in the Almaty region. The semi-arid Kazakh uplands lie to the north of Lake Balkhash, and the Saryesik-Atryan desert is just to the south. Three major streams feed Lake Balkhash, all from the south or south-east: the Ili River, the Karatal River and the Aqsu River. Since 1960, water levels in Lake Balkhash have been declining, mostly due to increased water usage for irrigation along the Ili and Karatal rivers.

Lake Balkhash faces many of the same problems as other water resources in the region. The water level has been severely reduced due to Soviet hydroelectric facilities, and this has seriously impacted upon the once-prominent fishing industry. The most obvious evidence of a decreasing lake level is the near-complete drying out of the reservoir on the southern part of Lake Balkhash. Only a small part remains, and that is likely to disappear soon unless remedial action is taken.

From 1972 to 2001, the southern part of the lake's surface decreased by approximately 15,000 hectares. Another sign of the diminishing water level is the newly formed island next to the south-east bank of the lake, and in other parts of the lake some islands have enlarged. Experts believe that in becoming more shallow and saline, Balkhash may have repercussions comparable to the tragedy of the Aral Sea.

The Balkhash system is heavily polluted by heavy metals and by chemical residues and fertiliser run-off from agriculture. The main water polluters are industrial, mining and refinery enterprises, livestock farms and irrigated farming. The Balkhash copper smelter heavily pollutes the lake with heavy metals and sulphites, and municipal wastewater treatment facilities are frequently overloaded or out of order. The over-exploitation of the Ili River has contributed to an increase in the salinity of water in the west of the country, and although Kazakhstan has strengthened its water management efforts in recent years, experts believe that with the population growth curve, agriculture, industry and urbanisation in the western areas of China, there is going to be more water use in the future. Thus there is a danger of a 'double whammy' effect developing.

As part of their industrial development plans, the Chinese government is settling many people from central China in the western Xinjiang Uyghur autonomous region. This population will maintain new industrial facilities and agricultural complexes that will also consume water, and thus put pressure on the rivers' ecosystems. The gradual degradation of the lake's ecosystems is being hastened by the construction of hydro-electric installations in China and following Chinese diversions of the Ili in the 1960s, for irrigation purposes.

The Irtysh River

The Irtysh River is the fifth longest river in the world and the largest in Asia in terms of its length and size of basin.

The Irtysh is considered one of the most polluted rivers in Eurasia due to its use by the East Kazakhstan and Pavlodar oblasts, where there are over 900 water-consuming enterprises, mainly related to metallurgy and mining. The main source of pollution is the drainage water from these industries, which is often discharged back into the river system. Since Soviet times, around 120 cubic metres of polluted drainage water is discharged to the rivers annually, or 60 per cent of total national discharge.

The situation around the Irtysh is already critical, and has been worsened since the flow from China to Kazakhstan has decreased. The decreased flow means that Kazakhstan is experiencing difficulty in operating several Irtysh hydropower facilities and port facilities, which are essential.

Due to its proximity to the industrial centres of Kazakhstan, China and Russia, the Irtysh has long been associated with river diversion talks. In the 1960s and 1970s, there were discussions based on river reversal schemes. While these schemes were not implemented, a smaller Irtysh-Karaganda canal was built between 1962 and 1974 to supply water to the dry Kazakh steppes and Kazakhstan's industrial Karaganda region. In 2002, pipelines were constructed to supply water from the canal to the Ishim River and Kazakhstan's capital, Astana. Additionally, in the 2000s, projects for diverting a significant amount of Irtysh water within China, such as the proposed Black Irtysh-Karamai Canal, have been decried by Kazakh and Russian environmentalists. The key problem concerning the Irtysh is that, after flowing through Siberia, it joins the Ob River and enters the Arctic Ocean. Therefore river pollution is not localised and will affect the global hydrological system.

It seems the lessons of history have been ignored. The war on nature, begun in earnest by Stalin, continues apace in the Ili-Balkash Basin and also in Baikonur and Saryshagan, where the Russians pay a handsome annual rent to Kazakhstan to enable them to pollute the environment with a constant stream of rocket launches.

CHAPTER FOURTEEN

Rocket Launching in Kazakhstan

There are currently 17 cosmodromes in the world which are used for launching rockets and space equipment, and these centres tend to cause considerable damage to the global climate and environment. The majority of global cosmodromes are situated on remote coastlines to reduce the danger of falling rocket stages in residential areas. Two cosmodromes, Baikonur Space Centre and the Saryshagan testing site, are located in Kazakhstan, which is one of only three countries to launch space equipment from continental sites (Russia and China being the other two). In doing so, the environment and health of the citizens is at risk due to the harmful effects of pollution and falling parts.

During the launch of a rocket from Baikonur, the majority of fuel is used for the initial propulsion up to a height of 40–50km while the rocket travels through the dense atmospheric strata. Approximately 500–600 tonnes of fuel are consumed during this phase, including around 200 tonnes of heptyl. Once this fuel is spent, redundant parts of the rocket are discarded. To accommodate the discarded parts, areas of the Karaganda, Akmol, Pavlodar and East Kazakhstan oblasts totalling approximately 35,000 square kilometres were designated as fall or 'dropping' zones.

These dropping zones also incorporate sections of the Nura, Ishim and Irtysh river basins and major cities such as Astana and Pavlodar. Additionally, many of the designated fall areas are covered by agricultural land, forests, reservoirs and even a large section of the Karkaralinsky National Nature Park, which covers around 1,000 square kilometres of pine and birch forests, hills and mountains.

The dropping zones span both Kazakh and Russian territory, and were initially chosen because they consisted of sparsely inhabited semi-deserts, deserts and tundra which were unproductive for agriculture. However, the landscape structure, fragility and uniqueness of such lands were not taken into account, and although they may be unproductive for economic purposes, these ecosystems have been detrimentally affected by the rocket-launching schemes.

Although specific areas have been designated for dropping zones, rocket fragments and discarded fuel containers have been found to deviate as far as 80–90km beyond the designated areas, and pollutants can be dispersed over thousands of square kilometres. Therefore even areas adjacent to the dropping zones, including residential areas, industry and power stations, are at risk. The risk factors include the transportation and storage of rocket fuel and its various components, the contamination of land, and the contamination of aquifers and the atmosphere with toxic rocket and other fuel components. These areas and their inhabitants are also exposed to danger from falling rocket parts containing unprocessed fuel, explosions in the air and on the ground, and, often as a direct result, devastating fires.

Baikonur Cosmodrome

Baikonur Cosmodrome is situated in the Kzyl-Orda Oblast, and was originally built by the Soviet Union in the late 1950s as the main focus of the Soviet space programme. The centre is currently leased to Russia, and is still at the forefront of international space exploration, as it has an important role in building the International Space Station.

Since it opened, Baikonur has helped launch over 2,000 rockets and continues to launch an average of 24–30 rockets annually using the propellant heptyl. Heptyl is a toxic substance which damages both human health and the environment. It is estimated that when the first stages of the rocket are detached, between 0.6 to 4 tonnes of unburned heptyl is spilt. Although much of this is burned up on the descent to earth,

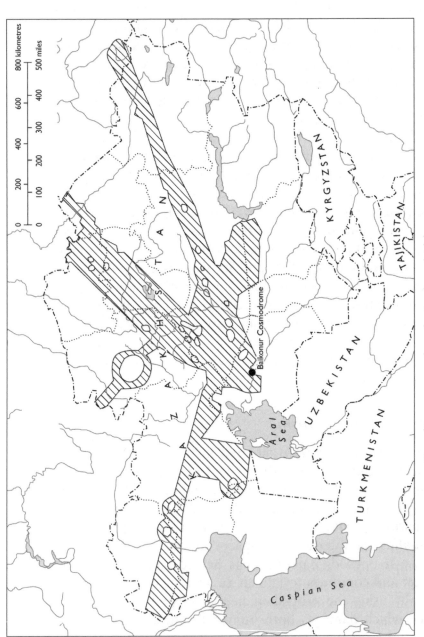

Landing areas of rocket's detachable parts.
(*Source: Environment and Development Nexus Kazakhstan 2004 –*
#UNDPKAZ06)

between 10kg and 30kg reaches the ground, where it is spread over the surface to evaporate or permeate the soil. Once it has penetrated the soil, heptyl is persistent and remains within the soil ecosystem for a considerable time. Scientists estimate that no fewer than 2,000 tonnes of heptyl have been spilled since Baikonur was established in Stalin's era.

Impact of Rocket Launching on the Environment

Launching rockets and space equipment from the Baikonur Cosmodrome has taken its toll on the natural environment and the atmosphere. Natural and man-made landscapes are being polluted with various hazardous chemical substances. The rocket launching and testing has had a particularly detrimental effect on Kazakhstan's atmosphere. Four common problems associated with the Baikonur Cosmodrome include the burning out of oxygen; the poisoning of the atmosphere by unburned fuel; trans-boundary migration of dust and partial depletion of the ozone layer; and the full-scale destruction of the ozone layer and creation of ozone 'holes'.

In addition to these factors, another negative impact is the development of dangerous meteorological phenomena as a result of rocket launching. Launching such equipment can instigate hurricane winds, snowstorms and heavy rain. Typically, a rocket launch can cause a decrease in air temperature, an intensification of the atmospheric front, an increase in atmospheric instability and the intensification of winds in the surface layer. Such effects cause heavy thunderstorms and rain a day after launching, sometimes exceeding norms by two to three times.

Rocket launches from Baikonur have also had a considerable impact on various Kazakh water resources when discarded parts and unburned fuel lands in river basins or reservoirs. Many pollutants used for the launches have limited solubility in water and can last for a long time, sometimes up to nine months in river waters, whereas in lake waters, heptyl, for example, resides for over eight years on average.

Contamination of soil is a particularly prominent problem,

as heptyl is very persistent, especially in dry soils in semi-arid regions such as Kazakhstan. Heptyl is very hazardous due to its volatility, and possesses the ability to accumulate and migrate in soils and silt. The moisture deficiency in the soil means that there is a lack of biodegradation of the compound, and there are no effective methods to neutralise heptyl. Therefore, lands contaminated by heptyl can remain contaminated for up 34 years.

In addition to the direct atmospheric, water and soil contamination from rocket parts, there is also a negative impact from the ground infrastructure of the cosmodrome and launch sites. The land and water resources around Baikonur Space Centre are contaminated with hazardous construction waste, scrap metal and spilled fuel.

Aside from efficient disposal of waste (as practised by any industry) and methods to address the environmental damage caused, there is not much that can be done to prevent the environmental problems caused by rocket launching in Kazakhstan. The only real solution would be to move the Baikonur Cosmodrome to a coastal site, but as this is a launching facility for the International Space Station, it is highly unlikely that this will happen. Further assessment into the ecological problems caused by rocket launching is needed, but this is difficult as the Kazakh environmental authorities do not have access to the space centre facilities and cannot really control the ecological situation in the launch area.

Trans-boundary Atmospheric Pollution in Central Asia

Atmospheric pollution is a serious environmental concern because contaminated air not only has a direct impact on humans, but it also serves as a transport medium that causes the pollution of soil cover, water sources, vegetation and foodstuffs. The quality of atmospheric air has undergone significant changes over the last 100 to 150 years in every region of the world, as the twentieth century was a period of active and unprecedented human interference in the state of our natural environments.

Although much of the trans-border pollution in Central Asia comes from trans-boundary waterways and land degradation, atmospheric pollution also has serious consequences. As is often the case, the poorest segments of society generally pay the greatest price for environmental mismanagement in terms of their sacrificed health and quality of life. Industrial zones in northern Kazakhstan and in the Uzbek portion of the Ferghana Valley are areas of concern where power plants which mostly burn coal and natural gas contribute significantly to urban air pollution, thus driving Central Asian carbon dioxide emissions per unit of GDP to among the highest in the world. This is threatening the region's industrial competitiveness and thus is of both local and global concern.

Atmospheric Pollution in Kazakhstan

The industrial regions of west and northern Kazakhstan are the areas which contribute most to atmospheric pollution in Central Asia. The Karaganda-Termirtay region is particularly problematic, and is now considered an environmental disaster area due to pollution from its coal mines, steel complexes, metallurgy enterprises and big power stations.

Within Kazakhstan, monitoring of ambient air pollution is carried out on a regular basis in 20 major cities and industrial centres. The negative impact on the environmental state of air is mainly caused by emissions and wastes from cities and industrial centres (stationary sources of pollution), vehicles (mobile sources of pollution), rocket and testing sites, forest and steppe fires and the flaring of gas and petroleum products at production sites.

The metallurgy industries are the major emitters contributing to air pollution in Kazakhstan but the oil and gas industry negatively affects the environment of west Kazakhstan. Extremely high levels of air pollution are registered, including nitric, sulphuric and carbon oxides, and such harmful emissions are caused mainly by low-quality coal used in power plants. A 2004 UNDP report suggested that, on average, 163kg per capita of various chemical compounds are released annually into Kazakhstan's atmosphere. In Karaganda Oblast this figure is 793kg, Pavlodar Oblast 547kg and 270kg in Atyrau Oblast. The energy sector is the main contributor of greenhouse gases and has contributed to climate change, which is another environmental concern in Kazakhstan. Over the past 100 years, the average temperature in the country has increased by 1.3°C, which is more than twice the global increase.

Global Consequences

Atmospheric pollution in Kazakhstan is not just a local and national problem; it is a regional and even a global environmental concern, as it affects communities and land outside its borders. For example, plumes emitted by forest fires (agricultural burning) in Kazakhstan have been observed from northern Alaska. Additionally, gaseous pollutants have been regularly observed in the Arctic since the 1950s. These are thought to be mostly of man-made origin due to emissions from Europe and Asia that are transported to and trapped in the Arctic air mass during the winter and early spring.

Mercury is also considered a global pollutant, and in its atmospheric form it has a long residence time in the atmosphere. It can be transported and deposited to remote places even 1,000km away from the source. Furthermore, mercury can be converted to methyl metrcury and accumulated in the food chain, posing a potential threat to human health. Mercury is released to the atmosphere from natural and anthropogenic sources such as coal combustion, waste incineration, metal mining, refining and manufacturing, and chlorine-alkali production. China heads the list of mercury emissions from anthropogenic sources, but Kazakhstan is close behind, with emissions stemming from large-scale mining activities.

The impact of air pollution on human health is complex, and quantification of the effect of any specific pollutant difficult, but medico-hygienic research carried out in Kazakhstan showed that the main reason for 80 per cent of morbidity cases and 20 per cent of mortality cases is most probably the emission of carbon monoxide, sulphur dioxide, nitrogen oxides and hydrocarbons, which all contribute to respiratory diseases.

When I first visited Almaty over a decade ago, the snow-capped mountains of the Tien Shan Range could always be clearly seen, rising majestically behind the city. Even in the heat of summer, the snow remains on the highest peaks, and it is possible to drive from the centre of the scorching city to the cool air of the mountain tops in less than an hour. Now, a green fug covers Almaty for most of the year, a visible sign of the increased traffic and increased pollution. It is a rare occurrence indeed for the mountains to appear. Within hours of arriving in Almaty, visitors can often develop a sore and rasping throat. Air pollution is rife.

Activities to Address Air-Pollution Issues

The most prominent measure to address atmospheric pollution in Kazakhstan is the 2002 Law 'On Protection of Atmospheric Air' which 'identifies concepts and principles of monitoring on the state of atmospheric air, standards for air quality and harmful physical effects on it and issues of licensing'. Nevertheless, the the United Nations Development Programme noted that mechanisms to reduce emission levels have not been wholly effective. One idea for protecting the environment that has already been tried is the application of green taxes on cars, trucks and factories. Unfortunately, this initiative has failed, mainly because such taxes are simply passed straight to the consumers through increased electricity, gas or service charges, proving to be no burden to the companies themselves and failing entirely to cut emissions. It is often found that the pollution charges are generally too low, so that producers prefer to pay fines rather than invest in abatement equipment and technology. The conclusion is that the present environmental protection system does not protect the public and the environment from severe exposure to air pollution.

CHAPTER FIFTEEN

Other Environmental Catastrophes in the Former USSR

It would be wrong to list the catalogue of environmental disasters perpetrated deliberately by Stalin and his successors in the Soviet Union without also mentioning some of the major environmental accidents that have blighted the USSR, as well as some of the more recent attempts at wrecking the ecosystem which seem to ignore warnings from the past.

K-19 *Submarine*

K-19 was one of the first two Soviet nuclear submarines equipped with nuclear ballistic missiles, specifically the R-13 missile. The vessel was launched on 11 October 1959, but due to the large number of accidents which had occurred during its construction, it had gained the unofficial nickname of 'Hiroshima' among naval crew.

On 4 July 1961, *K-19* was submerged off the coast of southern Greenland when it developed a major leak in its reactor coolant system, causing water pressure to plummet and the coolant pumps to fail. Coincidentally, a second accident disabled the long-range radio system, so they were unable to contact Moscow. The reactor temperature rose uncontrollably, reaching 800°C (1,470°F) and almost melted the fuel rods. Without coolant to control the reactor, the submarine's captain made a drastic decision and decided to install a makeshift cooling system by cutting off an air vent valve and welding a water-supplying pipe into it. The submarine's engineers were forced to work without anti-radiation suits or adequate protective clothing on board, meaning that they were sure to be lethally contaminated. They knew that they were effectively being sent to their certain deaths as they endeavoured to make the repairs.

The explosion contaminated the crew, the ship and some of the onboard weapons systems. The entire crew suffered severe radiation, and the seven engineers who repaired the damaged machinery all died of radiation exposure within a week. Twenty more died over the next few years. Instead of persevering further, the captain decided to head south to rendezvous with diesel submarines. The whole debacle prompted fears of a potential mutiny, and so the captain decided to have all small arms thrown overboard, only keeping a handful of small pistols for his closest officers. Eventually, a diesel submarine picked up *K-19*'s low-power distress transmissions and came to its rescue.

However, American warships in the same vicinity heard the call for help and even offered their assistance, which was a rare event during the Cold War. Nevertheless, the submarine's captain feared the Americans' motives and refused their help. Instead, he sailed to meet the *S-270*, which evacuated his crew and towed the submarine home. Here the damaged reactors were replaced, a process which took two years. During this time, there were further radiation leaks and localised poisoning of the environment and the engineers who conducted the repairs.

K-431 *Submarine*

The *K-431* was a Soviet submarine that had a nuclear reactor accident on 10 August 1985. An explosion occurred during refuelling at Chazhma Bay, Vladivostok. Although relatively unknown in the Western media, *Time* magazine identified the accident as one of the world's 'worst nuclear disasters'.

The submarine's two pressurised water reactors utilised enriched uranium as fuel. On 10 August, after refuelling, the lid of the reactor tank was being replaced but was installed incorrectly and was repositioned a second time with the control rods attached. A beam intended to aid the movement was also wrongly positioned, and the lid was lifted up too far. This caused a partial meltdown and an explosion of radioactive steam.

The explosion ejected the recently acquired fuel which destroyed the machine enclosures and ruptured the submarine's pressure hull and aft bulkhead. It also damaged the fuelling shack and sparked a serious fire which took four hours to extinguish. On assessing the radioactive contamination, it was found that the majority of radioactive debris fell within 50 to 100 metres of the submarine, but a cloud of radioactive gas and particulates blew to the north-west across a 6km stretch of the Dunay Peninsula, narrowly missing the town of Shkotovo, 26 miles from Vladivostok.

Lake Baikal

Lake Baikal is the second most voluminous lake on the planet, after the Caspian Sea, with an average depth of 744.4m. It contains roughly 20 per cent of the world's surface fresh water, and is located in the south of Siberia. It is home to many unique plants and animals.

Lake Baikal is also the deepest and among the clearest of all lakes in the world. At more than 25 million years of age, it is also the world's oldest lake. It is home to more than 1,700 species of plants and animals, two-thirds of which can be found nowhere else in the world. For this reason, it was declared a UNESCO World Heritage Site in 1996. It is also home to the Buryat tribes, who reside on the eastern side of the lake, where they rear goats, camels, cattle and sheep. Temperatures in the area never rise above 14°C in the summer, but can fall to as low as −19°C in winter.

There are a number of environmental concerns associated with Lake Baikal due to the industrial policies of the former Soviet Union. The Baykalsk Pulp and Paper Mill (BPPM) was constructed in 1966, directly on the shoreline. Here, waste chlorine from paper bleaching was discharged into Lake

Baikal. After decades of protest, the plant was finally closed in November 2008, though not due to the protests, but to unprofitability. In March 2009 the plant owner announced the paper mill would never reopen. However, on 4 January 2010 production was resumed. On 13 January 2010 Vladimir Putin introduced changes which legalised the operation of the mill and brought a wave of protests from ecologists and local residents. This was based on Putin's visual verification from a mini-submarine. 'I could see with my own eyes – and scientists can confirm – Baikal is in good condition and there is practically no pollution,' he claimed.

In 2010, the BBC reported that the UN may remove the world's deepest and oldest lake from the World Heritage list because of concerns over pollution by the pulp and paper mill.

Proposed Nuclear Plant

Lake Baikal is under further threat from a proposal by the Russian government to build the world's first International Uranium Enrichment Centre at an existing nuclear facility in Angarsk, 95km (59 miles) from the lake's shores. Critics believe that up to 90 per cent of the proposed plant's production would remain in permanent storage in the Lake Baikal region, with only around 10 per cent of the uranium-derived radioactive material exported to international customers. They claim that this would pose a further serious threat to the local environment.

CHAPTER SIXTEEN

The Lessons of History

George Santayana, the twentieth-century Spanish poet and philosopher, said: 'Those who cannot remember the past are condemned to repeat it.' This has often been bastardised into the more familiar misquotation: 'Those who fail to learn the lessons of history are doomed to repeat them.' Whatever way you choose to say it, sadly it is true, particularly when it comes to the way we treat the environment. During the Stalinist era in the USSR, the environment and nature were regarded as an enemy to be tamed and enslaved, much in the same way as the human population. As we have seen, the major projects undertaken by the Soviets had a catastrophic impact, the consequences of which are still being suffered today and may be with us for generations to come. The Aral Sea, Vozrozhdeniye Island, the Polygon and the nuclear dumps all had their origins in the megalomaniacal plans of Stalin.

Undeterred by these lessons, mankind has embarked on a new campaign of environmental vandalism, but this time it has been packaged and disguised as the war against climate change and global warming. By pretending to be engaged in an all-out effort to cut CO2 emissions and march towards a new 'Green Jerusalem', politicians and big industry together are now busily destroying the planet once again, wrecking biodiversity, uprooting the world's air-conditioning system and causing desertification.

As the measure of life forms within a given ecosystem, biodiversity provides an essential environmental service to the planet, and its benefits to society are manifold. Biodiversity directly influences many important facets of life, including air quality, climate, water purification, pollination and the prevention of erosion. But with the relentless quest to produce more and more food for our rapidly expanding population, coupled with the drive for biofuels, vast areas of land and the seabed are being destroyed for all productive purposes, creating new, barren deserts.

We have known for a long time the important role that our forests and peat bogs play in capturing and storing carbon. We call this green carbon. But now there is increasing awareness and attention being paid to the crucial role of our oceans and marine ecosystems in maintaining our climate. Over half of all the biological carbon captured in the world – around 55 per cent – is sequestered by marine living organisms in the sea, which is why it is called blue carbon. As greenhouse gas emissions increase, including from countries experiencing periods of rapid economic growth like China, Brazil and India, they cause ever greater impacts on worldwide weather patterns. We can see the massive effect this is having on food production and human lives and livelihoods. Every day we add a further 22 million metric tonnes of CO_2 to our oceans. That's why maintaining and improving the ability of our oceans to capture and store CO_2 is of vital importance to human survival.

That is also the reason we cannot afford any longer to overlook the critical role of our oceans. Without the essential ecosystem service they provide, climate change would be far worse. Recent research has indicated that a tiny part of the marine environment – the mangrove swamps, salt marshes and seagrasses that cover just 0.5 per cent of the seabed – account for the capture of at least half, and maybe three-quarters, of this blue carbon. They are our blue carbon sinks. Keeping them in good shape could be one of the most important things we can do to keep climate change under control.

But to achieve this we have to reduce the rate of marine and coastal ecosystem degradation. We need to extend the emissions trading system to embrace blue carbon. Carbon credits for marine and coastal ecosystem CO_2 capture and storage should be traded and dealt with in a similar way to green carbon. We need to establish a global blue carbon fund to pay for the protection and enhancement of remaining seagrass meadows, salt marshes and mangrove forests through effective management.

We need to improve energy efficiency in marine transport, including the fisheries, aquaculture and maritime tourism sectors. We need to encourage sustainable, environmentally sound ocean-based energy production, including algae and seaweed. Vast industrial schemes to cover swathes of our coasts and ocean floors with wave, tidal and offshore wind farms must be subject to intensive impact assessments, to ensure that they do not do further damage to marine ecosystem services.

Blue carbon lies at the very heart of the global warming debate. It is the key to climate change. Unfortunately, over the past 60 or 70 years we have lost around 20 per cent of the seagrass meadows, mangrove swamps and salt marshes that play this vital role in CO_2 reduction. That trend has to be reversed. Our survival depends on it.

Our insatiable thirst for natural materials, farmland, the mining of minerals and access to fresh water supplies means that man-made deforestation is rapidly destroying the world's oldest living ecosystem. There is no rainforest in the world which has been left undisturbed by humans, be it for wood, food, animal products or tourism.

Peatland is Europe's equivalent of rainforest, and it constitutes a vital component of the world's natural air-conditioning system. Peatland and wetland ecosystems accumulate plant material and rotting trees under saturated conditions to form layers of peat soil up to 20 metres thick, storing on average ten times more carbon per hectare than other ecosystems. As a natural carbon sink, peatland is present in 180 countries and covers 400 million hectares, or 3 per cent of the world's surface. Ironically, vast areas of carbon-capturing peat bogs in Europe are being torn up to make way for so-called 'greener' sources of energy, such as industrial wind turbines, rendering this whole process carbon-negative.

The construction of the mammoth steel towers, huge blades

and vast concrete foundations under every turbine, along with the quarries, borrow pits, drains, service ('floating') roads, overhead power lines and pylons means the carbon footprint from every industrial wind turbine built on deep peat often exceeds any environmental benefit it may bring. But such developments are being driven by a sea of subsidies. The big power companies are no longer farming wind, they are farming subsidies, and the poor consumers will have to foot the bill.

International experience to date has demonstrated that industrial wind power is unviable without heavy government subsidies and inflated feed-in tariffs. In every country where wind turbines have been installed they have failed to demonstrate economic feasibility, they have failed to demonstrate viability as a solution to global warming, they have failed to achieve significant CO_2 reduction, and they have failed to provide efficient electricity production or protection of the environment.

Indeed in countries where industrial wind power has been added to the grid in any volume, consumer costs have skyrocketed. The two countries with the highest numbers of installed commercial wind turbines in Europe – Germany and Denmark – now have the highest electricity bills in Europe. And yet in Germany, *Der Spiegel* reported in a recent article that despite 20,000 installed turbines, CO_2 emissions have not been reduced by even a single gram, because additional coal-burning plants have had to be built to support wind power.

In the UK, the introduction of destabilising wind energy to the grid has meant extensive resort to gas-burning facilities and greatly increased consumption of gas.

This in turn has dramatically driven up the price of gas to UK consumers. In Spain, a study by Juan Carlos University has laid the blame for the country's worsening economic crisis on the wind industry. The report states that the surging price of electricity has driven most of Spain's large energy consumers out of the country.

What we are witnessing is a dramatic transfer of money from the poor to the rich; from the beleaguered consumers to the wealthy estate and land owners and power companies, driving many people into fuel poverty. The most vulnerable people in society will be forced to make the choice between food or fuel.

And the environmental destruction associated with industrial wind turbines doesn't stop there. Wind turbines require powerful magnets as a key component in their generators. These magnets are made from neodymium, which is extracted from rare earth metals in an industrial process that creates huge toxic wastelands in parts of China, poisoning land, crops, animals and people.

The construction of industrial wind turbines has a direct impact on wildlife habitats, protected species of birds and ecologically fragile water courses. Turbines chop up rare birds, including white-tailed eagles in Norway and golden eagles in California. Their constant low frequency noise and vibrations are intolerable to livestock, wildlife and humans. There are innumerable reports of the long-term, irreversible and destructive impact that industrial turbines have on wildlife, leading to abandoned habitats, as well as the negative impact on livestock performance and production.

And of course, if animals suffer such consequences, it is easy to understand why humans are driven to despair when giant turbines are built near to their homes. Low-frequency sounds that travel easily and vary according to the wind constitute a permanent risk to people exposed to them. The effects of such broad-spectrum low-intensity noise, especially at night, combined with shadow flicker and vibration, which can affect individuals indoors as well as outdoors, have caused people to abandon their homes in distress and ill health. The profits being made by the landowners and power companies is at the expense of people's lives.

With the global race on to build mega wind turbines with a 275-metre wingspan, the time has come to pull the plug on this madness. Wind farms are environmentally destructive, economically foolish and aesthetically appalling. They will not cut carbon emissions and they will double or treble electricity bills. This wanton trashing of the countryside in pursuit of a flawed political dream would have brought a smile to the lips of Joe Stalin himself. It is a money-making scandal fuelled by greed and driven by politicians who are ignorant of the facts. It has got to be stopped.

Scientists say the extent of biodiversity loss worldwide is horrendous. They believe that between 150 and 200 species are being lost every 24 hours. Much of those losses can be attributed to climate change. We need to teach the public that biodiversity is valuable; it has an economic, social, aesthetic and practical value from which every one of us individually benefits. Ecosystem services purify the air we breathe, act as a global air-conditioning system, provide us with rainfall and oxygen, and fertilise plants. We have never put a price tag on these ecosystem services because they are invaluable. But sadly, like Stalin during Soviet times, some people think that anything that is free has no value and therefore can be exploited and abused.

Now we are learning that these things do have a cost and we are paying the price. Mankind's greatest ever contribution to the planet has sadly been a negative one – global warming. Thousands of years from now, if earth survives, the only remaining trace of our current civilisation will be our carbon footprint. Over the next quarter century, global energy consumption is forecast to grow by 61 per cent. Over two billion people still do not have access to any power at all.

If oil at over \$100 per barrel is painful for us, it is an impossible agony for developing countries. If we continue to rely on fossil fuels as our main energy source, we will exacerbate world poverty, face catastrophic increases in global temperatures, create freak weather conditions and cause sea levels to rise by over a metre, wiping out tens of millions of people worldwide.

We have to change. Reducing CO_2 emissions by 20 per cent in Europe by 2020 is a start, but it is not nearly enough. We need to aim for zero CO_2 emissions, and the technology is already here to achieve this goal. The current oil, gas and coal technologies are in their twilight years. They are sunset technologies. You need only look at the oil spill which occurred in the Gulf of Mexico in 2010 to see that drilling for oil in environmentally sensitive areas is completely untenable. That rules out oil exploration in the Arctic.

Even nuclear power, which is an almost CO2-emission-free energy producer, cannot provide the answer. It would require an estimated 200,000 new nuclear plants around the world to replace the base load energy currently provided by coal, oil and gas. The capital costs, security risks and unresolved question of nuclear waste storage render such a prospect inconceivable. The legacy of nuclear weapons tests in the Polygon of East Kazakhstan must act as a stark reminder of the consequences of misusing nuclear technology. Of course nuclear power will play a role in any CO2 emission-free future, but the role will be nominal rather than significant, particularly after the melt-down that caused widespread panic at Japan's Fukushima Daiichi nuclear plant following the March 2011 earthquake and tsunami.

Carbon is the enemy and reducing carbon emissions is the main task confronting the world. If we don't get it right, we will face a new challenge of environmental refugees – people fleeing drought and famine brought about by global warming. But we must not allow ourselves to be driven down the wrong paths by knee-jerk reactions and quick-fix subsidised projects.

We will have to work in partnership with countries outside the EU to encourage them to join in a sustainable, low-carbon future. A low-carbon economy can create new jobs and prosperity. On the other hand, the cost of doing nothing will be one order of magnitude greater than the cost of tackling climate change now. We need to invest a lot more resources into developing the hydrogen economy which, I believe, will be the next great industrial revolution. Hydrogen, which is the lightest and most abundant chemical element in the universe, can be readily stored and can provide an effective energy source. In Germany, they are building hydrogen-powered cars, trains and ferries. Hydrogen-powered homes are under construction. We need to cut our dependency on fossil fuels and look to the future.

I also believe that we can save 75 per cent of the energy we currently use by being more efficient. It is shocking that we still allow homes to be built in parts of Europe with single-glazed windows and no loft and cavity wall insulation. Triple glazing and proper insulation would cut our energy bills dramatically.

In the meantime, we have to take care that the policies we pursue are sustainable. The drive to produce biofuels is causing global deforestation, which, as well as releasing massive quantities of stored CO_2 into the atmosphere, could also lead directly to global famine. We are potentially creating a bigger global problem than we set out to resolve. In the US, enormous quantities of maize are being converted to bio-ethanol. This in turn has led to huge tracts of the Amazonian rainforest being burned to make way for growing maize and soya as food crops to make up the shortfall. It is claimed that the amount of maize required to convert into bio-ethanol to fill the tank of an average American family saloon would feed a human being for a whole year.

Meanwhile the Indonesian rainforest is being torn up to make way for biofuel crops like palm oil to supply the EU market. Such policies are destroying the world's air-conditioning system while at the same time releasing millions of tonnes of CO_2 into the atmosphere. In fact, deforestation is responsible for more greenhouse gas emissions than all the world's cars, trucks, planes and boats combined.

Echoing Stalinist policies, greed instead of care for the environment has become the defining feature of our strategy for tackling climate change, and the race to biofuels is potentially threatening the lives of millions of people as the global population soars from its present six billion to an estimated nine billion by 2050. An extra six million people are born every month. By 2030, the world population will have expanded by such an extent that we will require a 50 per cent increase in food production to meet anticipated demand. By 2080, global food production will need to double. But the reality is that an area the size of the Ukraine is being taken out of agricultural food production every year due to drought and as a direct consequence of climate change. Global food production is declining rather than expanding, and our headlong rush to produce biofuels is taking even more land out of food production.

The Disappearing Bees

A striking example of mankind's war on the environment is the issue of the disappearing bees. Many people think that bees are simply good for making honey. That bees provide us with honey is certainly true, and honey is a highly nutritious and wholly natural wonder-food. But bees are far more important than that. They are a vital part of our ecosystem and are essential to our survival. Almost 70 per cent of global food crops require pollination. Yet bees are dying out globally at an alarming rate. With no bees, we would be forced to live without products such as flowers, nuts, soya beans, onions, carrots, broccoli, sunflowers, apples, oranges and much, much more. Alfalfa, used for cattle feed, is also dependent on the honey bee; its loss would mean a fall in meat production. Certain medicinal plants and cotton rely on insect pollinators. Worldwide, 90 commercial crops require pollination to survive. If the bees die out, mankind might follow soon after.

The vanishing bee syndrome, or as it has become known, 'Colony Collapse Disorder' (CCD), started to filter through in news reports in 2007. For two years running, a third of all honeybees in the EU and US have mysteriously died. In some countries like Germany and Slovenia the losses have been as high as 60 per cent. Reports suggest the bees either completely disappear without trace or are seen crawling out of their hives to die.

There is already a part of China experiencing life without bees. The uncontrolled use of pesticides is said to have killed off the bees in an area of southern Sichuan in the 1980s. The pear trees, which are famous in this part of China, now have to be pollinated by hand, a laborious and costly process. Hundreds of villagers dip feathered sticks into pots of pollen and then have to dab each individual flower on each individual tree.

Pesticides were blamed in China, but what is causing the mass disappearance of bees in Europe? Scientists have yet to find a definitive answer. But they all seem to come back to the same conclusion. Bee colonies are stressed. Changing climate, poor air quality, monoculture and the overuse of some toxic chemicals have all taken their toll on bee health. A stressed or unhealthy bee is more susceptible to disease and to the increasingly prevalent varroa mite.

The *Varroa destructor* is an external parasitic mite that can only replicate in a honey bee colony and attacks honey bees causing a disease called varroatosis. The mite attaches itself to the body of the bee and weakens it by sucking haemolymph. In this process, the mite spreads diseases like deformed wing virus to the bee. A significant mite infestation will lead to the death of a honey bee colony, usually in the late autumn through early spring.

Treatment for varroatosis is extremely costly. The EU makes the regulatory approval of medicines so bureaucratic and expensive that suppliers have little option but to pass on the extra cost to the beekeepers. The more expensive these medicines become, the fewer beekeepers can afford to use them, allowing the disease to flourish. The whole situation has become so miserable that many beekeepers have simply withdrawn from this costly and unpredictable business.

Some people point the finger at pesticides, but this argument has been challenged even by beekeepers. Pesticides are by definition a toxic chemical. However, if their application to crops is strictly controlled, bees can still continue to thrive unharmed. Both bees and pesticides have lived together for a long time. There are beekeepers who suffer bee losses where no pesticides have been used in the area. There are also reports of bee deaths in the days before pesticides were ever used. However, if the bees become stressed for whatever reason, their resistance to the toxicity of pesticides or the parasite varroa is reduced.

The main culprit for our disappearing bee population seems to be monocultural farming, reminiscent of the vast cotton-covered plains of Uzbekistan. But instead of cotton, the former rich, green pastures full of nectar-producing plants have in some parts of the world become cereal deserts. Cereal crops like barley and wheat are pollinated by the wind and therefore bees don't feed on them. They are forced to look elsewhere for food. Without readily available food, the whole colony perishes. Bees confronted by vast swathes of cereals when they leave their hives become stressed and disease-prone.

Farmers in Canada have overcome this problem by planting beneficial crops in set-aside land in order to provide the valuable nectar that bees require. We could reward farmers for sowing small strips of land with crops such as the exotic flower phacelia, together with borage, charlock, wild white clover and other nectar-rich plants, which would create a haven not only for bees but for birds and other animals and insects. Small strips of land sown with these relatively cheap seeds would benefit not only bees but our whole ecosystem.

The loss of bees is not just a problem for beekeepers but for the whole world. Probably the most fundamental link in the food chain, the honey bee is fast becoming the weakest.

Emptying Our Oceans

We have also virtually succeeded in emptying our seas and oceans of fish. In our greed for endless resources, we have overexploited more than 80 per cent of global fish stocks, leavings some species, like Atlantic bluefin tuna and cod, teetering on the brink of total collapse. In Europe, under the Common Fisheries Policy (CFP), Brussels has imposed a ludicrously complex system of TACs (Total Allowable Catches), quotas and effort limitation, or restrictions on the number of days a fisherman can go to sea. Added to this is a welter of technical measures, mesh sizes, gear restrictions and even kilowatt days – which measures the output of a boat's engine and limits the number of days a fisherman can operate.

This bureaucratic obsession with micro-management has driven the industry to despair and yet, at the same time, it has done little to resolve the problem of diminishing fish stocks. On the contrary, the appalling obscenity of discards is a direct by-product of the current top-down system of management imposed by Brussels. Discarding fish has been forced upon our fishermen by the CFP. If a fisherman attempts to land an undersized or out-of-quota fish, he will be prosecuted and end up with a criminal conviction and a hefty fine. Rather than face criminal charges, our fishermen have no alternative other than to dump these fish dead over the side. An estimated 1.7 million tonnes of fish are discarded in this way every year in the EU, many of them in the North Sea. Such wanton waste and wholescale disregard for the environment and sustainability cannot be allowed to continue if mankind is to survive.

But far from curbing these excesses, we seem to turn a blind eye to illegal fishing, which is worth an estimated $23 billion every year. This vast criminal enterprise, often controlled by the Triad and the Mafia, steals fish on such a scale that some coastal stocks are now nearing collapse, threatening the survival of poor rural communities, often in developing nations.

Dilapidated and dangerous industrial trawlers using banned fishing gear remain at sea for months or sometimes years to avoid surveillance or inspection. They refuel at sea, are resupplied at sea and change over their crews at sea. They hoover up everything in their path, killing tens of thousands of seabirds, turtles, sharks and other protected species, and dumping millions of tonnes of by-catch. Their sole motivation is to maximise profits and minimise costs, and to achieve this they break all the rules, enslave their crews and sail under so-called 'flags of convenience' to avoid detection.

Flags of convenience are issued by countries like Honduras, Panama and the Grenadines. The authorities in these countries pocket a handsome profit by issuing licences to these pirate trawlers from China, Korea, Russia and elsewhere, allowing them to fly their flags while they steal fish.

Unscrupulous owners of these pirate vessels recruit crew by offering wages of $200 per month, low by our standards but irresistible to desperately poor people from coastal states in West Africa like Sierra Leone and Mauritania. These people then find themselves trapped on board a rusting hulk, their identification cards confiscated by the skipper, and told that their food and accommodation costs $250 per month, i.e. more than their wages. They are effectively enslaved.

The abuse of these slave workers is horrific. The Environmental Justice Foundation filmed one trawler off West Africa where the owners had erected a wooden hut hanging over the stern end of the ship. This was home to more than 200 crew forced to sleep, cook, eat and wash in conditions which would be frowned upon in the worst slums of Mumbai. The majority of workers are untrained, and yet are forced to work with dangerous and poorly maintained equipment in one of the world's most deadly professions. Workers who complain or try to escape are brutally beaten. There are even allegations of murder, with bodies flung into the sea. Many hundreds are seriously injured every year, but they have no protection. These boats are beyond the law.

The EU, the US and even the UN have a crucial role to play in tackling this global scandal. Some European boats are known to be engaged in pirate fishing, flying flags of convenience. Lists of these vessels have been given to the European Commission. They should be named and shamed. We also know that the big pirate trawlers operating off the coast of West Africa routinely transfer their catches at sea to smaller vessels, which mix the illegal fish with legally caught fish and then land the combined catch in ports of convenience like Las Palmas in Gran Canaria. This fish is then fully certified and sold into the European market, despite the fact that it has been caught and stored under these desperately unhygienic and unsanitary conditions. Many of the global pirate owners even have headquarters and offices in Gran Canaria from where they direct their worldwide criminal activities. They must be exposed, closed down and prosecuted.

The international community needs to get tough on pirate fishing. They need to crack down on those nations who tolerate this economic and environmental scandal. The United Nations needs to enforce a worldwide register of industrial fishing vessels. This would be relatively easy to do, and such enforced transparency would enable fishing authorities around the world to root out the pirates and put an end to this global crime.

Europe is now only 40 per cent self-sufficient in fish for human consumption. We import 60 per cent of all the fish we eat. Much of this imported fish comes from pirate sources. EU consumers are therefore financing this vast criminal project with all its attendant human-rights abuse, environmental damage and impoverishment of poor coastal communities.

Meanwhile, our own fishermen are forced to obey an over-abundance of rules and regulations conjured up by armies of bureaucrats in Brussels. But our fishermen obey the rules. It is sickening for them to see huge imports of illegally caught fish from outside the EU undercutting their markets. There must be zero tolerance for Illegal Unregulated and Unreported (IUU) fishing.

The Third Industrial Revolution

But I am an incurable optimist. After what I have seen I need to be positive. That is why I believe we are on the threshold of the Third Industrial Revolution, which, like the first two, will be driven by the convergence of new energy regimes with new communication regimes. The first hydraulic agricultural societies – Mesopotamia, Egypt, China, India – invented writing to manage the cultivation, storage and distribution of grain, which was used to feed slaves who, in turn, provided the energy to build the pyramids and cities and run the ancient economies.

In the modern era, the link between coal-powered steam technology and the printing press gave birth to the first industrial revolution. It would have been impossible to organise the dramatic increase in the pace, speed, flow, density and connectivity of economic activity made possible by the coal-fired steam engine using the ancient hieroglyphs and oral forms of communication.

This in turn was followed in the late nineteenth and throughout the first two-thirds of the twentieth century by first-generation forms of electrical communication: the telegraph, telephone, radio, television, electric typewriters, calculators, etc. These inventions, many of them Scottish, converged with the introduction of oil and the internal combustion engine, enabling the organisation and marketing of the second industrial revolution.

Now the smart technologies which gave us the internet and

vast distributed global communication networks will be used to reconfigure the world's power grids so that people can produce renewable energy and share it, people to people, in the same way they currently share information. The creation of a renewable energy regime, partly stored in the form of hydrogen and distributed via smart inter-grids, will open the door to a Third Industrial Revolution, transforming the way we live in the twenty-first century.

Renewable forms of energy – solar, micro-wind, hydro, geothermal, ocean waves and biomass – make up the first of the three pillars of the Third Industrial Revolution. Although only providing a tiny amount of the current global energy mix, these sunrise technologies are attracting massive amounts of private and public investment, as governments around the world mandate targets and benchmarks for their widespread introduction.

To maximise renewable energy and minimise its cost, it will be necessary to develop storage methods that facilitate the conversion of intermittent supplies of these energy sources into reliable assets. Renewable energy, by its very nature, is unreliable. When the wind isn't blowing, or the sun isn't shining, or the water isn't flowing because of drought, electricity can't be generated. But if we use some of the electricity generated while renewable energy is abundant to extract hydrogen from water, which can then be stored for later use, society will have a continuous supply.

When hydrogen is used as an energy source, the only by-products are pure water and heat. Our spaceships have been powered by high-tech hydrogen fuel cells for more than 30 years. Electricity produced from renewable sources can be used to split water into its separate elements of hydrogen and oxygen by the process of electrolysis. Hydrogen can also be extracted directly from energy crops, animal and forestry waste and organic rubbish – so-called biomass – without the need for the electrolysis process. By utilising hydrogen as a storage carrier for all forms of renewable energy, society can guarantee a constant supply of power. Of course the cost of hydrogen is still relatively high, but new technological breakthroughs and economies of scale are dramatically reducing costs year on year. Moreover, hydrogen-powered fuel cells are at least twice as efficient as the internal combustion engine.

The EU has become the first superpower to flag up the transition to the Third Industrial Revolution by making a binding commitment to producing 20 per cent of EU energy requirements from renewable sources by 2020. The European Commission has established a massive research initiative called the Hydrogen Technology Platform to develop further the hydrogen economy. Meanwhile Chancellor Angela Merkel of Germany has committed £350 million to hydrogen research in Germany and called for a Third Industrial Revolution in several speeches. Already hundreds of hydrogen-powered fuel cell forklifts, scooters, cars, buses and trucks are being used throughout the EU. A hydrogen fuel cell submarine is in operation in Germany and hydrogen-powered ferries are under development there and in the Netherlands. I was privileged to test drive a hydrogen-powered car in Strasbourg. It was smooth, highly responsive and fast.

So the first two pillars of the Third Industrial Revolution – a shift to renewable energy and an aggressive hydrogen fuel cell technology research and development programme – are now in place. The third pillar – the reconfiguration of the power grid along the lines of the internet, allowing businesses, industries, individuals and homeowners to produce and share their own energy with each other – is already being tested by power companies in Europe.

By fostering renewable energies, a hydrogen infrastructure and a continent-wide intelligent international grid, we can help create a sustainable economic development plan and turn the dream of an integrated single market into a reality for our 500 million EU citizens in the first half of the twenty-first century.

Greening the Desert

So what can we do to reverse the destruction of nature that mankind has wrought? How can we atone for the environmental crimes of Stalin and the Soviet Union? What steps can we take to arrest the spread of deserts and the loss of productive agricultural land? The answer may lie in permaculture. Permaculture is about designing ecological human habitats and food production systems. It is a land use and community building movement which strives for the harmonious integration of human dwellings, microclimate, annual and perennial plants, animals, soils, and water into stable, productive communities. The focus is not on these elements themselves, but rather on the relationships created among them, by the way we place them in the landscape. This synergy is further enhanced by mimicking patterns found in nature.

There was a recent fascinating example of this theory put into practice on the Jordanian desert, not far from the Dead Sea and only a few kilometres from the spot in the River Jordan where Jesus was baptised. When I visited this place in 2009, my Jordanian guide pointed to the tiny trickle of water that is all that is left of the once mighty River Jordan. 'It is a good job that Jesus was baptised 2,000 years ago,' he said. 'There wouldn't be enough water today to baptise him!' It is true. This is one of the most arid deserts in the world. The soil is completely dry. It has a high salt content. There is little rain from one year to the next. Herds of goats overgraze the only scraggy vegetation that struggles to survive, like giant maggots stripping the flesh from the bones of a rotting carcass.

Yet it was on this arid landscape that some pioneering agriculturalists demonstrated how permaculture can transform even the most unlikely soils. First they dug a long trench, around two metres wide and half a metre deep. They filled the trench with loads of mulch, taken from desert trash which the local farmers traditionally burn. They laid a network of micro-irrigation systems along each side of the ditch and then waited for the first rain to fall, jealously harvesting every drop. Soon the trench filled up and the water started to seep into the surrounding soil.

Next they planted nitrogen-fixing desert trees along one side of the ditch, to provide shade and to generate natural fertilisation. On the other side of the ditch they planted date palms, fig trees, guavas, pomegranates and even citrus trees. Within only four months the fig trees were already bearing fruit. Salt levels in the soil were gradually falling as the salt was locked into inert, insoluble compounds that could no longer harm the ecosystem. Soon, to the astonishment of local farmers, mushrooms could be seen growing in the soggy mulch of the trenches. They had never seen mushrooms before, and indeed they had never seen this level of moisture before in the desert.

Within less than a year, an area of 40 acres had been turned into a veritable Garden of Eden, where once had been only desert. And this had been achieved without huge expense, without artificial fertilisers and without chemicals and massive amounts of water. It was proof positive that permaculture could provide the answer to the world's problems of poverty and hunger.

A central theme in permaculture is the design of ecological landscapes that produce food. Emphasis is placed on multi-use plants, cultural practices such as sheet mulching and trellising, and the integration of animals to recycle nutrients and graze weeds. The Jordanian project embraced all of these key principles.

However, permaculture entails much more than just food production. Energy-efficient buildings, waste water treatment, recycling and land stewardship in general are other important components of permaculture. More recently, permaculture has expanded its purview to include economic and social structures that support the evolution and development of more permanent communities, such as eco-housing projects and eco-villages. As such, permaculture design concepts are applicable to urban as well as rural settings and are appropriate for single households as well as whole farms and villages.

Permaculture principles focus on thoughtful designs for small-scale intensive systems which are labour efficient and which use biological resources instead of fossil fuels. Designs stress ecological connections and closed energy and material loops. The core of permaculture is design and the working relationships and connections between all things. Each component in a system performs multiple functions, and each function is supported by many elements. The key to efficient design is observation and replication of natural ecosystems, where designers maximise diversity with polycultures, stress efficient energy planning for houses and settlements, using and accelerating natural plant succession and increasing the highly productive 'edge-zones' within the system.

In the broadest sense, permaculture refers to land use systems which promote stability in society, utilise resources in a sustainable way, and preserve wildlife habitat and the genetic diversity of wild and domestic plants and animals. It is a synthesis of ecology and geography, of observation and design. Permaculture involves ethics of earth care because the sustainable use of land cannot be separated from lifestyles and philosophical issues. Permaculture adopts techniques and principles from ecology, appropriate technology, sustainable agriculture and the wisdom of indigenous peoples. The ethical basis of permaculture rests on care of the earth – maintaining a system in which all life can thrive. This includes human access to resources and provisions, but not the accumulation of wealth, power or land beyond their needs.

We have a long way to go before we will see the deserts bloom again and the seas and oceans teem once more with fish. Mankind has cruelly abused this tiny planet on which we have a tentative foothold. No one exemplified this abuse more stridently than Joseph Stalin, who once famously said: 'One death is a tragedy. One million is a statistic.' Perhaps the legacy that he left the world will act as a sharp reminder that we are merely tenants of this temporary home and not its owners. We cannot wreck and ravage our environment and expect our children and grandchildren to inherit our catastrophes and deal with the consequences. Like any responsible tenants, we must care for our home and always seek to leave it in a better condition than that in which we found it. That would create a legacy future generations would be proud of.

Bibliography

Allouche, J. (2005). 'A source of regional tension in Central Asia: The case of water', *CP 6: The Illusions of Transition: which perspectives for Central Asia and the Caucasus?*' pp. 92–102

Allouche, J. (2007). 'The governance of Central Asian Waters: national interests versus regional cooperation', *Disarmament Forum*, 4: 45–55

Bennion, P., Hubbard, R., O'Hara, S., Wiggs, G., Wegerdt, J., Lewis, S., Small, I., van der Meer, J. and Upshur, R. (2007). 'The impact of airborne dust on respiratory health in children living in the Aral Sea region', *International Journal of Epidemiology*, 2007: 1–8

Bohr, A. (2003). 'Regional Cooperation in Central Asia: Mission Impossible?', Helsinki *Monitor, Quarterly on Security and Co-operation in Europe*, (Special Issue on Central Asia) (The Hague: Martinus Nijhoff Publishers), 14(3): 254–268

Bosch, K., Erdinger, L., Ingel, F., Khussainova, S., Utegenova, E., Bresgen, N. and Eckl, P.M. (2007). 'Evaluation of the toxicological properties of ground- and surface-water samples from the Aral Sea Basin', *Science of the Total Environment*', 374: 43–50

Carius, A., Feil, M. and Tazler, D. (2003). '*Addressing Environmental Risks in Central Asia: Risks, Policies, Capacities*', OSCE-UNEP-UNDP, Environmental Governance Series, Adelphi research

CNS Occasional Papers. 'Former Soviet Biological Weapons Facilities in Kazakhstan: Past, Present and Future' [online], available at: cns.miis.edu/opapers/op1/index.htm (accessed 25/01/10)

Crighton, E.J., Elliot, S.J., van der Meer, J., Small, I. and Upshur, R. (2003). 'Impacts of an environmental disaster on psychosocial health and well-being in Karakalpakstan', *Social Science & Medicine*, 56: 551–567

Crighton, E.J., Elliot, S.J., van der Meer, J., Small, I. and Upshur, R.

(2003). 'The Aral Sea disaster and self-rated health', *Health &*
Place, 9: 73–82

Dahl, C. and Kuralbayeva, K. (2001). 'Energy and the environment
in Kazakhstan', *Energy Policy*, 29: 429–440

Drobzhev, V.I., Gordienko, G.I., and Mukasheva, S.N. (2004). 'Iono-
sphere disturbances during rocket launches at Baikonur (Kazakhstan)', *Mathematics and Computers in Simulation*, 67: 433–439

Dukhovny, V., Mirzaev, N. and Sokolov, V. (2008). 'IWRM Implementation: Experiences with water sector reforms in Central
Asia'. In: Rahmann, M.M. and

Erdinger, E., Eckl, P., Ingel, F., Khussainova, S., Utegenova, E.,
Mann, V. and Gabrio, T. (2004). 'The Aral Sea disaster – human
biomonitoring of Hg, As, HCB, DDE and PCBs in children living
in Aralsk and Akchi, Kazakhstan', *International Journal of Hygiene and Environmental Health*, 207: 541–547

EU and CA. 'European Union and Central Asia, EU action on water
resources in Central Asia as a key element of environmental
protection', European Commission External Relations

Farmer, B. (2008). 'Afghanistan promotes pomegranates over opium
poppies in farming overhaul' [online], available at: http://www.te-
legraph.co.uk/news/worldnews/asia/afghanistan/3491421/Afgha-
nistan-promotes-pomegranates-over-opium-poppies-in-farming-
overhaul.html (accessed 22/02/10)

Froebrich, J. and Kayumov, O. (2004). 'Water Management aspects
of Amu Darya'. In: Nihoul, J.C.J., Zavialov, P.O. and Micklin,
P.P. (ed). *'Dying and Dead Seas – Climatic Vs. Anthropogenic
Causes'*, NATO Science series: IV, Earth and Environmental
Sciences, (36), Kluwer Academic Publisher

Funakawa, S., Suzuki, R., Karbozova, E., Kosaki, T. and Ishida, N.
(2000). 'Salt-affected soils under rice-based irrigation agriculture
in southern Kazakhstan', *Geoderma*, 97: 61–85

Geist, H.J. and Lambin, E.F. (2004). 'Dynamic Casual Patterns of
Desertification', *Bioscience*, 54(9): 817–828

Guterstam, B. (2008). 'Towards sustainable water resources management in Central Asia'. In: Rahmann, M.M. and Varis, O. (ed).
*'Central Asian Waters: Social, Economic, Environmental and
Governance puzzle'*, Water and Development Publications: University of Helsinki Technology

Hagt, E. (2003). 'China's water policies: Implications for Xinjang
and Kazakhstan' [online], available at: www.cacianalyst.org/
?q=node/1358_ (accessed 26/01/2010)

Hansen, J. and Galieva, E. (2005). 'Final Evaluation of UNDP Semipalatinsk Programme Outcomes', UNDP

Hazlewood, P. (2009). 'Pomegranate trees line Afghanistan's road to prosperity' [online], available at: http://www.telegraph.co.uk/expat/expatnews/6384851/Pomegranate-trees-line-Afghanistans-road-to-prosperity.html (accessed 22/02/10)

Heaven, S., Ilyschuchenko, M.A., Kamberov, I.M., Politikov, M.I., Tanton, T.W., Ullrich, S.M. and Yanin, E.P. (2000). 'Mercury in the River Nura and its floodplain, Central Kazakhstan: I. River sediments and water', *The Science of the Total Environment*, 260: 35–44

Heaven, S., Ilyschuchenk o, M.A., Kamberov, I.M., Politikov, M.I., Tanton, T.W., Ullrich, S.M. and Yanin, E.P. (2000). 'Mercury in the River Nura and its floodplain, Central Kazakhstan: II. Floodplain soils and riverbank silt deposits', *The Science of the Total Environment*, 260: 45–55

Huskey, E. (1997). 'Kyrgyzstan: A case study for potential conflict', *The Soviet and Post-Soviet Review*, 24(3): 229–249

ICG (2002). 'Central Asia: Water and Conflict', Asia Report No. 34

ICG (2002). 'Central Asia : Border Disputes and Conflict', Asia Report No. 33

ICG (2005). 'The Curse of Cotton: Central Asia's destructive monoculture', Asia Report No. 93

ICG (2006). 'Central Asia: What Role for the European Union', Asia Report No. 113

Iskakov, M. and Tabyshalieva, A. (2002). 'Cold Winters Upstream, Dry Summers downstream in Central Asia' [online], available at: www.cacianalyst.org/newsite/newsite/?+node/152_ (accessed 19/12/2009)

Jenson, S., Mazhitova, Z. And Zetterstrom, R. (1997). 'Environmental pollution and child health in the Aral Sea region in Kazakhstan', *The Science of the Total Environment*, 206: 187–193

Kajumov, A.K. and Mahmadaliev, B.U. (2002). '*Climate Change and its influence on public health*', Dushanbe 2002

Kangur, K. (2008). 'Deliberative water policy-making in Kazakhstan and Kyrgyzstan: Focus groups in the Talas and Chu River Basins'. In: Rahmann, M.M. and Varis, O. (ed). '*Central Asian Waters: Social, Economic, Environmental and Governance puzzle*', Water and Development Publications: University of Helsinki Technology

Karimov, A., Qadir, M., Noble, A., Vyshpolsky, F. and Anzelm, K. (2009). 'Development of Magnesium-Dominant Soils Under Irrigated Agriculture in Southern Kazakhstan', *Pedosphere*, 19(3): 331–343

Kemelova, D. and Zhalkubaev, G. (2003). 'Water, Conflict and Regional Security in Central Asia Revisited', *N.Y.U. Environmental Journal*, 1: 479–502

Khamzayeva, A. (2009). 'Water resources management in Central Asia: security implications and prospects for regional cooperation'. In: Khamzayeva, A., Rahimov, S., Islamov, U., Maksudov, F., Maksudova, and Sakiev, B. (ed). *'Water Resources Management in Central Asia: Regional and International Issues at Stake'*, Documentos CIDOB Asia 25

Li, P., Feng, X.B., Qiu, G.L., Shang, Z.G. and Li, Z.G. (2009). 'Mercury Pollution in Asia: A review of the contaminated sites', *Journal of Hazardous Materials*, 168: 591–601

Libert, B. (2008). 'Water Management in Central Asia and the activities of UNECE'. In: Rahmann, M.M. and Varis, O. (ed). *'Central Asian Waters: Social, Economic, Environmental and Governance puzzle'*, Water and Development Publications: University of Helsinki Technology

McCauley, D.S. (2005). 'Environmental Management in Independent Central Asia'. In: Burghart, D.L. and Sabonis-Helf, T. (ed). *'In the Tracks of Tamerlane: Central Asia's Path to the 21st Century'*, Center for Technology and National Security Policy (CTNSP) at the National Defense University

Megoran, N. (2004). 'The critical geopolitics of the Uzbekistan-Kyrgyzstan Ferghana Valley boundary dispute, 1999–2000', *Political Geography*, 23: 731–764.

Micklin, P. (2000). *'Managing Water in Central Asia'*, London: Royal Institute of International Affairs

Micklin, P. (2007). 'The Aral Sea Disaster', *Annual Review of Earth and Planet Science'*, 35: 47–72

Mikhalev, V. and Reimov, A. (2008). 'Land Degradation in Central Asia', *Energy and Environment*, 09/2008 [online], available at: http://www.developmentandtransition.net/index.cfm?module=ActiveWeb&page=Webpage&DocumentID=673_ (accessed 18/01/2010)

Moore, M.J., Mitrofanov, I.V., Valentini, S.S., Volkov, V.V., Kurbskiy, A.V., Zhimbey, E.N., Eglinton, L.B. and Stegeman, J.J. (2003). 'Cytochrome P4501A expression, chemical contaminants and histopathology in roach, goby and sturgeon and chemical contaminants in sediments from the Caspian Sea, Lake Balkhash and the Ily River Delta, Kazakhstan', *Marine Pollution Bulletin*, 46: 107–119

Mosello, B. (2008). 'Water in Central Asia: A prospect of Conflict or Cooperation?', Journal of Public and International Affairs, 15: 151–173

Nalwalk, K. (2000). 'The Aral Sea Crisis: The Intersection of Economic Loss and Environmental Degradation' [online], Available at: www.pitt.edu/~eofarb/aral2.pdf_ (accessed 25/01/10)

NBCentral Asia (2007). 'Trans-Afghan Powerline Seen as Worth the Risk' [online], available at: http://tdworld.com/overhead_transmission/trans-afghan-powerline/_ (accessed 02/02/2010)

Netalieva, I., Wessler, J. and Heijman, W. (2005). 'Health costs caused by oil extraction air emissions and the benefits from abatement: the case of Kazakhstan', *Energy Policy*, 33: 1169–1177

OECD (2005). '*Water and violent Conflict*', Development Assistance Committee, Issues Brief. Orlovsky, L., Orlovsky, N. and Durdyev, A. (2005). 'Dust storms in Turkmenistan', *Journal of Arid Environments*, 60: 83–97

OSCE-UNESCAP-UNECE (2006). 'Support for the Creation of a Transboundary Water Commission on the Chu and Talas Rivers between Kazakhstan and Kyrgyzstan', Final Project Report

Peachey, E.J. (2004). 'The Aral Sea Basin Crisis and Sustainable Water Resource Management in Central Asia', *Journal of Public and International Affairs*, 15: 1–20

Peyrouse, S. (2007). 'The Hydroelectric Sector in Central Asia and the growing role of China', *China and Eurasia Forum Quarterly*, 5(2): 131–148

Postel, S.L. and Wolf, A.T. (2001). 'Dehydrating Conflict', *Foreign Policy*, 126: 60–67

Rahimov, S. (2009).' Impacts of climate change on water resources in Central Asia', Water Resources Management in Central Asia: Regional and International Issues at stake, Documentos CIDOB Asia 25: 33–55

Roffey, R. and Westerdahl, K.S. (2000). 'Conversion of former biological weapons facilities in Kazakhstan: A visit to Stepnogorsk', Swedish Defence Research Agency

Saigal, S. (2003). '*Kazakhstan: Issues and Approaches to Combat Desertification*', Asian Development Bank/The Global Mechanism

Saiko, T.A. and Zonn, I.S. (2000). 'Irrigation expansion and dynamics of desertification in the Circum-Aral region of Central Asia;, *Applied Geography*, 20: 349–367

Sapozhnikov, D. G. and Tsvetkov, A. I. (1959). 'Precipitation of hydrous calcium carbonate on the bottom of Lake Issyk-Kul', *Doklady Akademii Nauk SSSR_*, 24: 131–133

Savvaitova, K.A. and *P12_175*P12_175Petr, T. (1999). 'Fish and Fisheries in Lake Issyk-Kul (Tien Shan), River Chu and Pamir Lakes' [online], available at: http://www.fao.org/docrep/003/X2614E/x2614e10.htm (accessed 10/02/10)

Sevcik, M. (2003). 'Uranium Tailings in Kyrgyzstan: Catalyst for Cooperation and Confidence Building?', *The Non-Proliferation Review*, Spring 2003

Seversky, I.V. (2006). '*Present and predictable changes of snow and glaciations of zone of flow formation and their possible influence on water resources of Central Asia*', Almaty 2006

Seversky, I.V. and Tokmagambetov, T.G. (2004). '*Current glaciations degradation of mountains of the Southeast Kazakhstan*', Almaty 2004

Sherman,S. and Rafikov, V. (1992). 'Sedimentation of the Nurek Reservoir', *Power, Technology and Engineering*, 25(10): 668–673

Sidorov, O. (2003). 'Central Asia's water resources as a cause of regional conflicts', *Journal of Social and Political Studies* [online], available at: _ http://www.ca-c.org/online/2003/journal_eng/cac-05/19.sideng.shtml (accessed 26/01/2010)

Sievers, E.W. (2002). 'Water, Conflict and Regional Security in Central Asia', NYU Environmental Law Journal, 10: 384–93

Sojamo, S. (2008). 'Illustrating Co-existing conflict and cooperation in the Aral Sea Basin with Twins Approach'. In: Rahmann, M.M. and Varis, O. (ed). '*Central Asian Waters: Social, Economic, Environmental and Governance puzzle*', Water and Development Publications: University of Helsinki Technology

Small, I., van der Meer, J. and Upshur, R.E.G. (2001). 'Acting on an Environmental Health Disaster: The Case of the Aral Sea', *Environmental Health Perspectives*, 109: 547–549

Spoor, M. and Krutov, A. (2003). 'The 'Power of Water' in a divided Central Asia', *Perspectives on Global Development and Technology*, 2(3/4): 593–614

Stevenson, S.J.S. (2006). 'Crying Forever: A Nuclear Diary', Dog Ear Publications

Stulina, G. and Sektimenko, V. (2004). 'The change in soil cover on the exposed bed of the Aral Sea', *Journal of Marine Systems*, 47: 121–125

Szayana, T.S. (2003). 'Potential for Ethnic Conflict in the Caspian Region'. In: Oliker, O. And Szayna, T.S. (ed). '*Faultlines of conflict in Central Asia and the south Caucasus: implications for the U.S. Army*', RAND: Arroyo Center

Tolosa, I., de Mora, S., Sheikholeslami, M.R., Villeneuve, J.P., Bartocci, J. and Cattini, C. (2004). 'Aliphatic and aromatic hydrocarbons in coastal Caspian Sea sediments', *Marine Pollution Bulletin*, 48: 44–60

Tookey, D.L. (2007). 'The environment, security and regional co-operation in Central Asia', *Communist and Post-Communist Studies*, 40: 191–208

UNDP (2004). 'Environment and Development Nexus in Kazakhstan: Publication in support of the Millenium Development Goals', Almaty, Kazakhstan: UNDP

UNDP (2004a). 'Water Resources of Kazakhstan in the New Millennium, Publication in support of the Millenium Development Goals', Almaty, Kazakhstan: UNDP

UNDP (2008). 'Review of Donor Assistance in the Aral Sea Region (1995–2005). Tashkent

(UNDP 2010). 'The High-Level International Forum "Uranium Tailings: Local Problems, Regional Consequences, Global Solution" took place in Geneva on 29 June 2009' [online], available at: http://www.un.org.kg/en/news-center/news-releases/article/65-news-center/3556-the-high-level-international-forum-uranium-tailings-local-problems-regional-consequences-global-solution-took-place-in-geneva-on-29-june-2009 (accessed 9/03/2010)

UNECE (2008). '2nd Environmental Performance Review: Kazakhstan', New York, UN

USAID (2009). 'Accelerating regional economic growth to provide licit alternatives to poppy production' [online], available at: http://afghanistan.usaid.gov/en/Program.20a.aspx (accessed 22/02/2010)

Varis, O. And Rahaman, R.R. (2008). 'The Aral Sea keeps drying out but is Central Asia short of water?'. In: Rahmann, M.M. and Varis, O. (ed). '*Central Asian Waters: Social, Economic, Environmental and Governance puzzle*', Water and Development Publications: University of Helsinki Technology

Villain, J. (1996). 'A Brief History of Baikonur', *Space Policy*, 12(2): 129–134.

Warneke, C., Bahreini, R., Brioude, J., Brock, C.A., de Gouw, J.A., Fahey, D.W., Froyd, K.D., Holloway, J.S., Middlebrook, A., Miller, L., Montzka, S., Murphy, D.M., Peischl, J., Ryerson, T.B., Schwarz, J.P., Spackman, J.R. and Veres, P. (2009). 'Biomass burning in Siberia and Kazakhstan as an important source for haze over the Alaskan Arctic in April 2008', *Geophysical Research Letters*, 36: 1–6

Weigerich, K. (2006). 'Illicit water: unaccounted but paid for', paper presented at CERES-IWE research seminar, 8 November 2006

Wegerich, K. (2008). 'Passing over the conflict; The Chu Talas Basin Agreement as a model for Central Asia'. In: Rahmann, M.M. and Varis, O. (ed). '*Central Asian Waters: Social, Economic, Environmental and Governance puzzle*', Water and Development Publications: University of Helsinki Technology

Wegerich, K., Olsson, O., Froebrich, J. (2007). 'Reliving the past in a changed environment: Hydropower ambitions, opportunities and constraints in Tajikistan', *Energy Policy*, 35: 3815–3825

Weinthal, E. and Luong, P.J. (2002). 'Environmental NGOs in Kazakhstan: Democratic goals and non-democratic outcomes', *Europe-Asia Studies*, 51(7): 1267–1284

Whish-Wilson, P. (2002). 'The Aral Sea environmental health crisis', *Journal of Rural and Remote Environmental Health*, 1(2): 29–34

Williams, J. (2007). 'Uzbekistan – the sick man of Central Asia', *Pesticide News*, 75: 3–4

Zetterstom, R. (1999). 'Child Health and environmental pollution in the Aral Sea region in Kazakhstan', *Acta Paediatr Suppl*, 429: 49–54

Index